The Undersea Discoveries
of Jacques-Yves Cousteau

THREE ADVENTURES
Galápagos, Titicaca, The Blue Holes

The Undersea Discoveries
of Jacques-Yves Cousteau

THREE ADVENTURES
Galápagos,
Titicaca,
The Blue Holes

Jacques-Yves Cousteau
and Philippe Diolé

Translated from the French by J. F. Bernard

Doubleday & Company, Inc.
Garden City, New York

ISBN: 0-385-06921-9
Translated from the French by J. F. Bernard
Library of Congress Catalog Card Number 72-93396
Copyright © 1973 by Jacques-Yves Cousteau
All Rights Reserved
Printed in the Federal Republic of Germany
9 8 7 6 5 4 3 2

CONTENTS

The Galápagos archipelago and its location in the Pacific with respect to Central America.

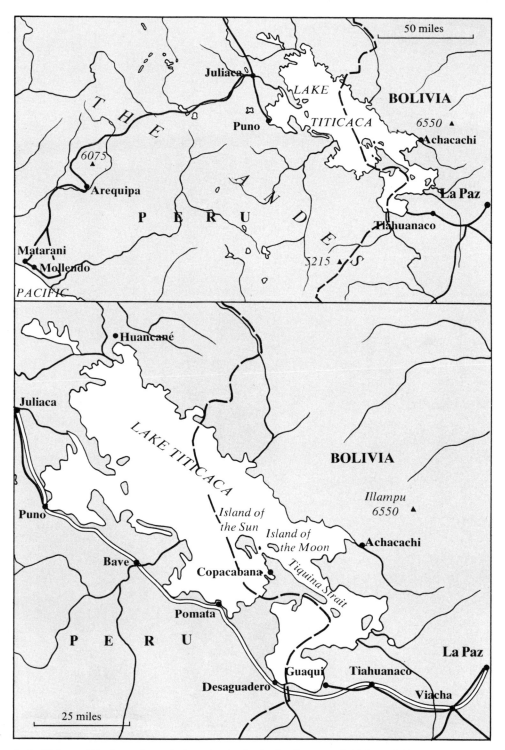

Lake Titicaca. The inset shows its location in the Andes, between Bolivia and Peru.

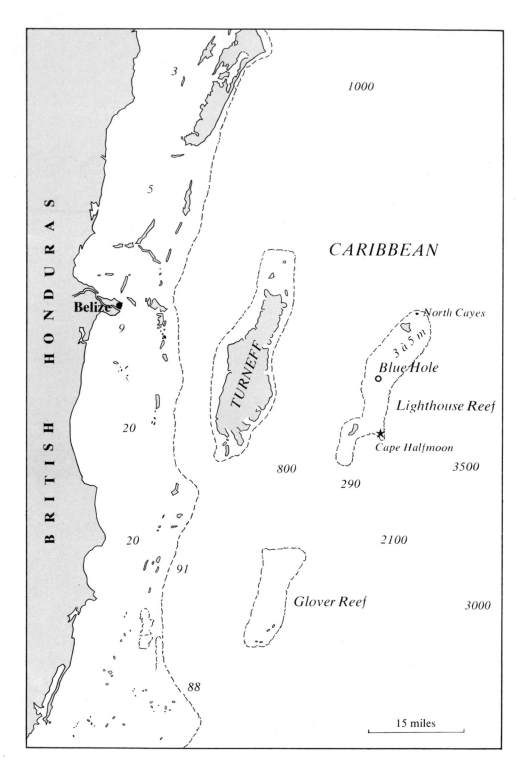

Lighthouse Reef, off British Honduras.

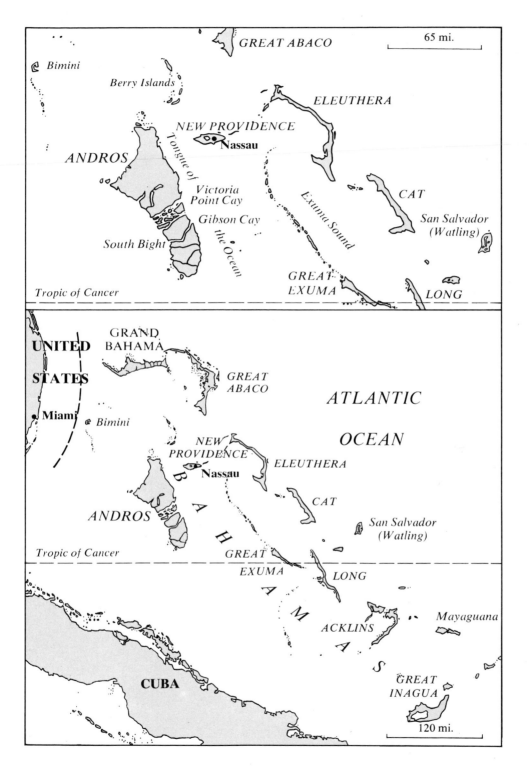

Above: Detail of the Bahamas, showing Andros Island. Below: The Bahamas.

PART ONE
Galápagos

CHAPTER ONE

In Darwin's Wake

The red and green dragon rose on its front paws. We could see the spiked crest running from its head to the tip of its tail. It was looking at us, but gave the impression of not seeing us. There was no sign of fear, and, although surrounded by six or seven men, it did not attempt to escape. Even when Yves Omer stretched out his hand to try and touch it, the animal only turned around quickly, took a few hurried steps, and then stopped and looked at us. It seemed to regard us more with indifference, or perhaps wariness, than with fear. It was sunning itself, and it would allow itself to be disturbed only for a good reason. Human beings did not appear to merit more attention or consideration than other species with which it lived — the turtles and penguins, and the sea lions which thought it a great joke to pull the dragon's tail while it was bathing.

This dragon is a reptile, a marine iguana—the only species of iguana which swims, feeds, and is perfectly at home in the water. And it is found in only one place in the world. The iguana is the main reason why *Calypso* is here, five hundred miles off the coast of South America, below the equator in the Galápagos archipelago. We wanted to see this animal, to film it, and to dive with it.

Jacques-Yves Cousteau giving instructions to the divers aboard *Calypso*.

The first meeting between the iguana and me was rather formal. *Calypso*'s team had preceded me to Galápagos by several days, and, when I arrived, everyone delighted in presenting their new friend to me and telling me what they had learned.

Unfortunately, I cannot always be with our divers, or aboard *Calypso*. As much as I would like to, I cannot take part personally in every episode of our films. I must divide my time, as best I can, between Los Angeles, Paris, New York, Marseilles, and Monaco. In many respects, this is a very inconvenient arrangement. Nonetheless, it has its consolations. It is a marvelous experience suddenly to be able to join our team and to listen while everyone communicates the excitement they have lived in the preceding weeks — the adventures with whales, hammerhead sharks, and — today — with marine iguanas.

On this occasion, at *Calypso*'s camp on Hood Island, I was introduced to the iguanas; and the things I was told about them were even more astonishing than if I had been with the divers as they observed the dragons of Galápagos and discovered some of the secrets.

Hood Island is a volcanic island. Viewed from the surface of the sea, it seems a black dry rock. But higher up, there is vegetation in truly tropical profusion. Like most islands of the archipelago, Hood has a great vertical wall on one side and, on the other, a more gradual slope into the sea. Seen on the black lava of Hood, the iguana takes on the appearance of a red and green monster-in-miniature — a being right out of the past. But it is a small monster, not more than four feet long. My idea was to ask Jacques Renoir, our cameraman, to film the iguana from close up, with a wide-angle lens, working from the head to the tail. This would be the opening sequence of our film; and the viewer would have the impression of seeing a living, terrifying dinosaur — whereas the iguana is, in fact, a completely inoffensive animal whose patience is inexhaustible in allowing itself to be observed, filmed, and approached.

Calypso's team is so expert that it is quite capable of collecting the necessary documents, organizing its work, and establishing a filming program without me. When I arrive, I usually do nothing more than suggest a new slant or throw out a few ideas which are immediately incorporated into our plan and put into practice. My role in such cases is to review and approve what has been done in my absence and to contribute my own thoughts.

Hood Island was the first base camp at Galápagos, and seven men stayed there. There were two cameramen, Michel Deloire and Jacques Renoir, and an assistant, Henri Alliet; a chief diver, Bernard Delemotte; two divers, Dorado and Delcoutère; and our American cook, "Little Joe."

From Arctic to Equator

Calypso had just spent three months filming sea otters in the Arctic. We caught them caring for their young, hiding among large algae (kelp), and breaking open clam shells with stones held against their chests. They are fascinating animals, but, like so many other species, sea otters even today are threatened with extinction. Their fur is the most expensive in the world; and, for that reason, they are in constant danger of being exterminated, despite all the steps that have been taken to protect them.

We also virtually completed a film on salmon in the Arctic. We had followed their migrations in the course of their upstream voyage in the rivers of Kodiak Island — the habitat of the largest bears in the world, giants which reach a length of over twelve feet and which fish for salmon with their paws.

The chill of the arctic autumn finally made us leave the Bering Straits and the Aleutians, and we set our course for the south. It seemed to me the proper time to undertake an expedition — of which I had long been dreaming — to the most exotic islands of the Pacific, the Galápagos archipelago. We were literally following in the wake of Charles Darwin, and, like him, we would find riddles aplenty to solve. Darwin visited the Galápagos Islands, aboard the *Beagle,* in 1835, when he was twenty-six years old.* The observations he made there on both finches and tortoises provided him with the first elements of his theory of the origin of species.

Coming as we did from the Arctic, the Galápagos seemed to us an equatorial paradise — enchanted isles where, far from the domain of man, strange and marvelous animals lived side by side in peace. We found marine iguanas, diving champions of the reptile kingdom; wingless cormorants, and giant turtles hundreds of years old.

Our journey to the Galápagos took us along the coast of California. En route, our cameramen took the opportunity to shoot several excellent sequences on manta rays. They encountered veritable schools of these animals, in which there were individual specimens measuring up to eighteen feet across.

After a stop at Acapulco on the coast of Mexico, *Calypso* set a course directly for Galápagos — a journey of about two days. The first islands sighted in the archipelago were Darwin, Genovesa, Bindloe, and Pinta, jutting out of the water, their summits obscured by fog and clouds. This initial view of the Galápagos gave the impression of bleakness. We were to discover, little by little, that the islands were not precisely what we had envi-

*See Appendix I.

A "booby" on the volcanic beach of one of the Galápagos Islands.

sioned when we abandoned the icy chill of the Arctic. For the moment, however, they seemed overrun by dense tropical vegetation. They resembled the isles of the Caribbean which we knew so well — Puerto Rico, or St. Barthélemy. Only one of the islands we saw actually lived up to our expectations: Pinta, which is a paradise of orchids.

It was off Darwin Island that we first sent out a Zodiac with a team of divers. The sea was rough that day, as though warning us that our dreams of tropical tranquillity were doomed to disappointment. The warning was not altogether necessary. As soon as our team disembarked on Darwin, they discovered the truth for themselves. Before them was a sheer cliff over three hundred feet high. It looked more like a paradise for mountain climbers than for divers. Nonetheless, our men set to work with a will, diving among the waves; and soon, Bernard Delemotte brought back the first marine iguana of the expedition.

So far so good. Delemotte, however, did not know quite what to make of his success. Animals of this kind were new in our experience, and Bernard was uncertain of the iguana's reaction to this intrusion upon its freedom. Would it bite? As it happened, the iguana's resistance was confined to wiggling, and Bernard was able to bring it aboard *Calypso* without mishap. Immediately, everyone gathered around the miniature dragon, more amazed by its prehis-

(Right) François Dorado carries a marine iguana back to the beach on Hood Island.

toric appearance than wary of its strangeness. What was most striking about the animal, it seems, was the length of its claws, which grow to about a third of the length of its fingers. The paws of the iguana are not prehensile; that is, they cannot grasp and hold.

Twenty Hammerhead Sharks

The second dive took place near Wolf Island. Yves Omer went down, and Bernard Delemotte accompanied him in the Zodiac. Shortly, Yves was back on the surface, shouting "Hammerheads! Hammerheads!" Delemotte immediately jumped into the water.

There were at least twenty hammerhead sharks swimming in tight formation. It was the first time in *Calypso*'s history that we had ever encountered these creatures in such a large school. We have no idea what circumstances had brought so many specimens together in this instance. They seemed merely to be passing through, between Wolf Island and the neighboring isle, where the current was particularly strong. Yves and Bernard, after swimming along with the sharks for a while and following them toward the open water in order to try for a few photographs, returned to the Zodiac and then rejoined their friends aboard *Calypso*. It had been an exciting and unexpected adventure; and, in addition to the sharks, they had sighted numerous giant sea turtles grazing on the bottom. But it was the hammerheads which had fired their imagination. "They were from twelve to fifteen feet long," Yves Omer said, "and the way in which they move makes them look even stranger than they are. It seems that a hammerhead shark tries to look all around itself by constantly moving its head back and forth, as though it were not enough that its eyes are located on those growths on each side of its head."

Our divers had often observed hammerhead sharks in the Red Sea and in the Indian Ocean. But this was the first time that anyone from our team had become involved with such a large group of them; and everyone aboard *Calypso* could easily understand the mixed feeling with which Yves and Bernard must have viewed these creatures. Hammerhead sharks are like something out of a nightmare. Everything about them is strange, unearthly. Their heads are flat and disproportionately wide, and their eyes are globular. The entire body seems alien and somehow horrifying. And the hammerhead's reputation does not belie its appearance. It is regarded as a potentially fearsome adversary and as the fastest and most agile — as well as the most unpredictable — of the sharks.

There are different opinions concerning the advantages and disadvantages to the hammerhead of its curiously flattened head. It is possible that it serves as a rudder and helps in turning sharply — hence the animal's agility. But it would be surprising if its eyes, located as they are at each end of the "hammer," function with greater efficiency because of their position. This has led some observers to conclude that sight is not the primary sense of the hammerhead and that numerous nerve cells, spread over the animal's body, play the principal role in the shark's perception of its environment.

These first dives, even though they brought *Calypso*'s men face to face with such formidably dangerous creatures, were more than worth while. They revealed that the waters of Galápagos contained a great wealth of marine fauna, and even — as we would discover when we landed on the island of Pinta — of botanical curiosities. There, our men would pass their leisure time hunting the innumerable orchids hidden in the moss of the larger trees.

Perhaps the most extraordinary thing about Galápagos is the striking contrast between the exuberant flora, the tropical humidity of some parts of the islands, and the surrounding waters, chilled by the Humboldt Current, in which we found fauna reminiscent of the arctic regions from which we had so recently come. It is not difficult to imagine our astonishment when, in an area that we thought of as "tropical," we ran into guillemots, exactly like those that we had seen a short time before in the Bering Straits.

We were delighted to be here, at the heart of this mysterious archipelago; but we were soon to discover that there were other islands of the group which, although also washed by the same cold currents, were less hospitable and fertile. For the moment, however, our minds were occupied by other matters. We had just crossed the equator, and it was time for us to proceed to the ceremonies which, in accordance with the traditions of the sea, must accompany such a crossing: the baptism of novices.

The Crossing

According to maritime practice, the ceremonies of the crossing should take place at the same latitude as the equator itself. Ours therefore were held before we arrived at San Cristóbal and Santa Cruz — at precisely the equatorial latitude. We began by installing a pool on *Calypso*'s forward deck. This was to be the scene of the formal initiation of those who were crossing the equator for the first time.

As is usual, the dignitaries who presided over the ceremonies were Amphitritos (our friend Yves de Pimodan) and Neptune (Bernard Delemotte). Under their aegis, the initiation proceeded strictly in accordance with the an-

The traditional ceremony of the crossing of the equator aboard *Calypso*. Amphitritos (Yves de Pimodan) and great Neptune (Bernard Delemotte) are in the center.

cient rubrics. First, a comic list of the sins, faults, and shortcomings of the novices was read publicly. Then Neptune and Amphitritos conferred together to determine the guilt or innocence of the accused and to weigh mercy against justice. The punishments to be meted out were announced by Dr. Blanc, who had exercised more than a little ingenuity in devising a sanction to fit every crime. The condemned, for example, were given a last meal of éclairs stuffed with cotton; then they were smeared with mustard, as though great Neptune regarded them as delectable frankfurters. When the god discovered his mistake, amends were made by cleansing the novices inside and out: they were given enemas and, as the final rite in the initiation, were thrown into the pool.

As in all initiations, there are occasional accidents in the ceremony of the crossing of the line. On this occasion, the victim was our engineer. When it came time for him to be thrown into the pool, rather than submit in humility he insisted in diving in under his own power. The trouble was that the pool had been constructed simply by draping a canvas on the inside of a shark cage—and the unruly initiate banged his head against the iron bars of the cage and lost a piece of scalp. He was stitched up by Dr. Blanc, who passed effortlessly from the role of judge to that of healer.

The truth is that *Calypso*'s men are rather expert at the conduct of these rituals. Our old-timers have conducted these ceremonies on several occasions—in the Indian Ocean, the Atlantic, and the Pacific.

(Left) The two merciless executors of Neptune's judgments. (Right) One of Neptune's victims, hoping to escape, plunges into our makeshift pool. It was a leap from the frying pan into the fire, for in the plunge he tore off a piece of his scalp.

Divers on Horseback

At noon on February 2, 1970, *Calypso* dropped anchor off the Island of San Cristóbal. San Cristóbal is the administrative center of the archipelago; and, at the port of Bahia Wreck, government officials visited *Calypso*. In the afternoon of the same day, our captain, Jean-Paul Bassaget, returned the courtesy.

A party went ashore to visit a statue of Darwin—the only one in the entire archipelago. The men stayed long enough to ascertain that San Cristóbal is not precisely what we mean when we say "tropical paradise." At sea level, the island is bare and arid, and its natural features are dead trees and black rocks. Here and there, prickly cactus grows, reminding one of the American Southwest. It is only at higher altitudes that there is a humid zone—forests of euphorbias, manchineel trees, and acacias. Altogether, it is a jungle of extraordinary beauty. Occasionally, one finds patches of land under cultivation. There are even fruit orchards.

One of our teams decided to explore these forests on the mountain slopes, and rented horses for the excursion. These were small, lively animals, and very nervous. Along with dogs, cows, and deer, horses were only recently introduced into the Galápagos archipelago, and they have already reverted to a half-wild state and do considerable damage to the flora of the islands.

(Right) "Darwin and the Origin of Species." (A caricature published in *Horizon* magazine and reproduced in Caravel Books, Cassell, 1968).

(Facing page) Owen Stanley's engraving of the *Beagle*. (Copyright, National Maritime Museum, Greenwich.)

(Following page) Two marine iguanas warming themselves in the sun.

The ride up into the mountains was less harrowing for our divers than the descent. The horses were so eager to get home that they refused to go down the slope at anything less than a trot. Our men therefore had to dismount and allow the animals to proceed at their own pace. The owners of the horses, however, who were serving as guides, were determined, although they were on foot, never to lose sight of their property. They therefore insisted on matching their pace to that of the horses—even when the animals broke into a gallop.

A Souvenir of Darwin

Calypso weighed anchor at San Cristóbal the following day, February 3, having obtained all the official permits and authorizations necessary for an exploration of the islands of the archipelago. There are many such islands— six large ones, twelve smaller ones, and forty very small ones. They are all volcanic in origin and contain a total of more than two thousand craters of various sizes. The volcanoes are not all dead in the Galápagos Islands. There was an eruption in 1968, on the island of Fernandina.

Calypso's first port of call was supposed to be Santa Cruz, where the Darwin Foundation is located. The Foundation is responsible for all research and study of the flora and fauna of the Galápagos, and we were interested in

Owen Stanley

obtaining, from the specialists of the Foundation, information that might be useful in selecting the most favorable sites for the filming that we had planned. Their advice seemed particularly important in view of the fact that, until *Calypso*'s arrival, there had never been any underwater filming in the Galápagos vicinity.

On our way to Santa Cruz, we stopped briefly at the small island of Santa Fe (also called Barrington Island); then we proceeded to Santa Cruz, where we dropped anchor in Academy Bay. The Darwin Foundation is located alongside the bay; and behind it, the heights of the island are often hidden in the clouds. These summits are no doubt covered with a wealth of interesting vegetation.

The director of the Foundation, Dr. Perry, welcomed *Calypso*'s men with the elegance and quiet cordiality which I, in my own mind, regard as typical of the British everywhere. He was even smoking a pipe.

The Charles Darwin Foundation was established in 1959, under the auspices of UNESCO. It has two primary duties. One, to study the animals of the archipelago; the other, to protect this "laboratory of evolution" in which one of the rarest animals in the world is still found.

Darwin visited the Galápagos Islands almost a century and a half ago. On September 17, 1838, his ship, H.M.S. *Beagle*, dropped anchor off the small island which the British call Chatham, and the Ecuadorians call San Cristóbal. The *Beagle* had sailed from England on December 27, 1831, on a

hydrographic mission; and Charles Robert Darwin, a young man of twenty-two, was the ship's naturalist.

Darwin, shortly before sailing, had taken his bachelor's degree. He had first intended to study medicine, but then theology had claimed his attention. But he never lost his predilection for the natural sciences. Although his background was one of staunch and militant Protestantism, it was not a conformist or limited Protestantism. It is possible, and even probable, that, before boarding the *Beagle,* Darwin already sensed that the stability of animal and vegetable species as set forth in the Bible and as taught in the schools did not correspond to the reality of nature. But, scrupulous theologian that he was, he insisted in corroborating his theories by observations in nature. From that standpoint, therefore, the mission of the *Beagle* was a heaven-sent opportunity for Darwin.

The ports visited by the ship along the coasts of South America had allowed the young naturalist to make observations which provided the elements of the theories taking shape in his mind. But it was in the Galápagos Islands that Darwin was to discover that which would convince him finally of the possibility of variation and adaptation among plants and animals of the same species. "The facts concerning the species of this archipelago," he wrote, "are at the basis of all my opinions."

Two species were most important to Darwin in confirming his theories. The first was the iguana, whose survival, bizarre appearance, and varying colors were bound to attract the attention of an observer as curious and careful as he. The second was the finch — a species which, from island to island, shows a diversity in form or in beak development corresponding to differences in their habits and especially in their nourishment. In other words, the finches of the Galápagos had adapted to local conditions. Some of the finches fed on grain. Others, Darwin observed, ate cactus. Over the centuries, these differences had resulted in the appearance of varieties of finches best adapted to survive in the conditions dominant on different islands of the archipelago. This, for Darwin, was the confirmation he had sought for his thesis on "the evolution of species."

In order to understand the complexity of the fauna of the Galápagos, it is necessary to recall that the archipelago occupies a very special location off the Pacific coast of South America. This chain of volcanic islands, in fact, is at the confluence of the warm currents from the North (especially that known as "El Niño") and of the cold current from the southeast — the Humboldt Current. This mixture of warm and cold waters has had a considerable impact upon the flora and fauna of the islands.

The fauna represents a strange mixture of tropical and polar species.

There are penguins and albatrosses, sea lions, and seals — all of which we are accustomed to find, not in the tropics, but in the Far North where we were filming salmon and otters. In the midst of these cold-water animals, we find iguanas, turtles, and even warm-water snakes.

The variations of temperature can have disagreeable effects — such as the dense fog which is not only inconvenient, but dangerous. But it is also responsible for the exceptional abundance of plankton in the waters surrounding the archipelago. And, thanks especially to the Humboldt Current, the marine fauna is extremely plentiful.

Armed with our cameras and with all the technological know-how and equipment of *Calypso*, as well as with a team of superlative divers, we came to the Galápagos with means of which Darwin never dared dream. Our purpose in coming, however, was the same as that of our eminent predecessor: to observe the phenomena which he had observed and, if possible, to experience the thrill and the astonishment which he must have felt in his own time.

Giant Tortoises

Calypso spent two days at anchor at Santa Cruz, where Dr. Perry and his assistant, with inexhaustible patience, briefed our divers on every aspect of marine life in the waters of the Galápagos which might possibly have been of value in our filming.

This layover at Santa Cruz also allowed some members of our team to observe the giant tortoises which are being bred systematically by the Foundation in the hope both of preserving the species and also of increasing its numbers. The variety of each individual island is bred separately from the others; and the offspring of the breeding process are returned to their island of origin. The differences among these varieties are so marked — at least to the experienced observer — that Governor Lawson informed Darwin, upon the occasion of his visit, that he was able to tell what island a tortoise came from simply by looking at it. The fact is that an evolutionary process was at work among these reptiles analogous to that of the finches. The shells of the giant tortoises were modified according to the kind of food that they hunted in their particular environment. But Darwin noted that they also differed according to the climate peculiar to each island: tortoises with very large domed shells inhabited the humid islands, whereas those with smaller shells were found in the more arid regions. And, he added, the shells were more or less curved in around the necks of the tortoises according to whether the reptiles ate grass or fed by tearing leaves from bushes, especially from crotons.

Until the early nineteenth century, these tortoises were plentiful in the Galápagos. Today, they have almost disappeared. There are several causes for this phenomenon. First of all, for more than two centuries the tortoises were hunted by buccaneers, privateers, and then by whalers. They would take the animals alive and keep them in the holds of their ships as a source of fresh meat; for a tortoise can survive for a year without food or water. Moreover, the fat of an adult tortoise, it was discovered, can render three gallons of oil. All that it was necessary to do in order to obtain this oil was to make a small incision near the tortoise's tail. Thus, between 1811 and 1883, some sixteen thousand of these animals were captured in the Galápagos.

There are other causes also. Goats, which had been put ashore by seamen, ate all the vegetation within reach, and thus deprived the tortoises of the grass and leaves on which they feed. Finally, volcanic disturbances opened up crevices in the islands, and many tortoises fell into these and, being unable to climb out, starved to death. In some holes, between twenty-five and fifty tortoise shells have been found.

The wealth of information obtained from the experts of the Darwin Foundation was not confined to giant tortoises. These gentlemen showed *Calypso*'s team a beautiful small beach, on the side of the island on which Punta Tamaya is located, where they were able to observe a numerous colony of gray iguanas. The weather was ideal on that occasion, the water was clear, and underwater visibility was about seventy-five feet. Even so, this second encounter between our team and the iguanas was marked by the same reserve as the first. The divers had no idea of how the iguanas would react. They did not wish to frighten them or disturb them. As it happened, they need not have worried. If our divers had never before observed iguanas in the water, it was equally true that the iguanas had never before encountered man in the water, and they gave every indication of being more surprised than frightened.

Across the Archipelago

The Darwin Foundation had been kind enough to show *Calypso*'s team a film which enabled our men to work out a tentative list of the places which were suitable for the project they had in mind. But it was still necessary to visit each of the larger islands individually, so as to ascertain the quantity and variety of iguanas offered by each.

The major problem of Jean-Paul Bassaget, *Calypso*'s captain, was that of finding supplies of food and water. For this, unfortunately, he would have to look elsewhere; for the Galápagos did not have much surplus to offer along these lines. The islands have no resources to speak of, not even fresh water.

The giant tortoises of the Galápagos are no longer found in great numbers, but they are now a totally protected species.

This commodity must be brought in by tanker — and even then there is not enough for the inhabitants, let alone for such visitors as *Calypso*'s men. Bassaget therefore had no choice but to make for Guayaquil, the major port of Ecuador, to take on water.

Before leaving, however, he intended to make sure that the divers were in the best possible situation, near a spot where the sea bottom was rocky and rich in algae. For this, it was necessary to spend a week sailing from island to island, in search of the right place.

Here are a few extracts from *Calypso*'s log:

Thursday, February 5. We leave Santa Cruz at five o'clock in the morning, with the intention of reaching Española, or Hood Island.

Passing within view of several islands, we get the impression that the volcanoes have not been inactive for very long. The traces of eruptions seem quite fresh. It seems to us that there has not even been enough time since the last eruptions for erosion to have set in. The lava streams are very well defined, and nothing has yet been worn away by the sea, the wind, or the sand. These signs of recent volcanic activity would also explain why there are no large trees on these islands — only a few mangroves and some hedges.

So long as it is on our way, we have decided to explore MacGowen Reef. We drop anchor nearby, in fifteen fathoms, and send out two teams. One of them will explore the reef itself, and the other will take a look at the bottom and report on underwater visibility.

The latter team sends back word that the reef ends in a vertical cliff; which means that *Calypso* can move in right alongside the reef, almost touching it. This might be a good spot, in fact, to set up a camp. Bernard Delemotte, who went down to 150 feet, has brought back some red algae which seem phosphorescent. He says that he could see them shining on the bottom.

At 9:25 A.M., with both teams back aboard, we set sail for Española. Less than two hours later, *Calypso* drops anchor off Hood Island, at Point Suarez.

Bassaget immediately sends out three teams: one will head for land in a Zodiac, and the other two, in launches, will spend the day exploring the surrounding bottom. The cameramen will be responsible especially for selecting sites suitable for filming and for reporting on the abundance of fauna.

By the end of the day, the three teams are back. Their reports are all encouraging, but not decisive. There are algae in quantity, but they are not nearly so photogenic as the California kelp, or the laminaria of Brittany's waters. On the other side of the ledger, all three teams report the presence, both on land and in the water, of many iguanas and of sea lions. The latter, they say, did not seem at all frightened by the sight of the divers.

One of the problems is that the water is very rough and not at all ideal for diving and filming. So far as setting up a camp is concerned, the news is even worse: the island is made of black lava. There is only a narrow beach, and it is covered at high tide.

If worse comes to worst, we will have to leave the divers here; but we should not do so until we are satisfied there is not a more favorable site on another island. In order to lose as little time as possible, we will send a team ashore so that we may begin filming. For this, however, we must choose a spot that more or less meets the conditions required for filming; that is, there must be iguanas, clear water, and an interesting marine background. So far as the condition of the water is concerned, we have very little to go on, since there has never before been any diving in the archipelago. We will therefore have to explore these islands carefully, but as quickly as possible, in order to find precisely the right location.

Calypso weighs anchor at 6:40 P.M. and, a short while later, we are alongside the island of Tortuga. Since it is clear and calm, the captain drops anchor and sends a reconnaissance team ashore. They report back that Tortuga is rocky and covered with lava, and that it is, if anything, worse than Hood Island. *Calypso* therefore gets under way again and, at six in the morning, reaches Isabela Island, one of the larger pieces of land in the archipelago. Passing through the Bolivar pass, we drop anchor in a protected spot at Point Espinosa off the island of Fernandina.

Once more we send out a reconnaissance team. They report back that, here too, conditions are hardly ideal. There are a great many iguanas — dark gray ones this time — and colonies of friendly sea lions and penguins. But the divers are unanimous in affirming that the water is too rough for underwater shooting. It would be a waste of time even to attempt it.

Calypso has not yet managed to find the proper site for a camp; and we are running low on food and water. Time is now of the essence, and we decide to return to Hood Island, which, now that we have seen other places, does not seem so bad. We start out at 8:00 P.M. — but almost immediately we are in the middle of a squall, with strong gusts and heavy rain. We must be extremely careful of the currents and of the rocky points that rise here and there above the surface.

By midnight, the water is calm again, and the *Calypso* is on her way to Hood Island.

Friday, February 6. At 9:30 in the morning, we drop anchor once more off Point Suarez. Bernard Delemotte and Michel Deloire immediately climb into a Zodiac and set off southward on a tour around the whole island in the hope of finding a place for a camp. But by now the sea is very rough, and some of the waves are twelve feet high. One entire side of the island is unapproachable because of the bordering cliffs which rise from 150 to 200 feet above the water. Against these vertical heights, the waves break with violence, and the spray seems to rise as high as the cliffs themselves. Delemotte and Deloire saw red and green iguanas in the water, but it was almost impossible to film them because of the tremendous swell which, combined with the rocky shore, made landing impossible along the island's southern shore. The iguanas, however, did not seem to mind and were quite at home among the rocks and waves.

Our divers also explored Gardner Bay, as well as the edges of a small island in the center of the bay which has a bright red color. The island is surrounded by underwater chasms in the sandy bottom; and these canyons are frequented by sharks.

Delemotte and Deloire, as the result of their reconnaissance, have decided that the only place we could land — a place inhabited by a sufficiently large number of iguanas for our film — is on one of the sides of Point Suarez, where there is a small bay protected from the violence of the waves by a line of black rocks. We can reach this bay by Zodiac, and there we should find relatively calm water. By some stroke of luck, there are two beaches along the bay, one of which is of fair size even though surrounded by lava cliffs. This strip of sand is about sixty feet long, narrow at one end and wider at the other; and it ends in dark, firm soil which is apparently beyond the reach of the breakers.

Female sea lions are capable of showing tenderness.

(Facing page) Two marine iguanas cross the black beach without disturbing a sleeping sea lion.

We have therefore decided to set up a camp at that spot. The team for the camp — seven carefully chosen men — is quickly put ashore. There are two cameramen, Michel Deloire and Jacques Renoir; Henri Alliet, an assistant cameraman; Bernard Delemotte, our chief diver; and two divers, Dorado and Delcoutère. And, as an entirely necessary part of the team, there is Little Joe, our American cook.

The following day is spent in setting up the camp and in unloading supplies — supplies and material. *Calypso*, now rationed to the maximum, leaves to its camera crew all that she can in the line of food and water. Then she weighs anchor for Guayaquil.

Friends in Camp

The campsite at Point Suarez, despite its black lava rocks, the violence of the waves and the aridity of the environs, proved in the final analysis to be well worth while. There was a large number of animals in the neighborhood, and they all seemed perfectly willing to be featured in a movie.

In the course of their exploration of the area, Delemotte and Deloire did not notice that the spot they chose for the camp was the long-established haunt of a troop of sea lions. The arrival of our team, however, did not seem to upset these animals in the least. Not only did they retain their former lifestyle intact, but they also accepted the newcomers to their neighborhood in good humor and struck up a very friendly relationship with them. For as long as our camera team was camped there, the sea lions made a habit of sleeping against the walls of the tent. During the hottest part of the day, one of the animals usually slept in the shade under the dining table. During lunch and dinner, the cameramen ordinarily used the sleeping sea lion as a footrest — a convenience which the animal seemed happy to provide.

The friendliness of the sea lions, so far as we could determine, had no selfish motives. They never once accepted food from the cameramen — not even a living fish. Apparently, all they asked of the relationship was to be allowed to mingle with the *Calypso*'s men and especially to be allowed to swim with them frequently.

Unique Individuals

The principal subject of the film planned by Michel Deloire and Jacques Renoir was the marine iguana. The filming site had been chosen with this in mind. We knew that iguanas live in groups which sometimes number as many as a thousand individuals; and the group on Hood Island was certainly that large. But in order for filming to proceed smoothly, it was not enough that there be a sufficient number of iguanas. It was also necessary that man and reptile become somewhat familiar with each other. After all, we know very little about iguanas, their reactions and their habits. In no place in the world, other than the Galápagos Islands, are there iguanas which swim, dive, and live on marine algae.

The iguana, at first sight, is not a very attractive animal. Even Darwin, who was sensitive to all animal life, describes iguanas as "being of hideous aspect, having a dirty black color, stupid, and sluggish in their movements." This, as we will see, is not an opinion entirely shared by *Calypso*'s men.

At the beginning of our sojourn in the Galápagos, all we knew about the marine iguana — *Amblyrhynchus cristatus* — was that it is usually between two and four feet long and that its tail accounts for at least half of that measurement.

We also knew, of course, that the appearance of marine iguanas varies from one island to another. For the iguanas which live on one island spend their whole lives there. They do not cross the stretches of water which separate the various islands of the archipelago. Therefore, there has been no interbreeding among the iguanas of different islands — a fact which has had the effect of preserving the original varieties. There are, in fact, eight subspecies of marine iguana. The reptiles of Hood Island, for example, are striped in red and green; on Cristóbal, they are dark gray; and at Santa Cruz, bronze green.

The first job of our team was to learn as much as it could about these living fossils. They did their job well, and, by the time that I myself reached the Galápagos Islands, they were able to pass on to me — with great enthusiasm — a good deal of new information on these miniature water dragons. For, by then, they had gotten to know the iguana very well indeed.

Calypso in the open water off Hood Island.

CHAPTER TWO

The Camp on Hood Island

After erecting their tent on the larger of the two beaches, well above the high-water line, *Calypso*'s team set up their kitchen, found a suitably dry spot for their compressor, and finally set up a small machine shop. They were then ready to take a look around the island on which they would be living for the next few weeks. Above them were huge rocks on which grew low, thin patches of vegetation: bushes and plants capable of living in a soil which is both volcanic and salty.

The island, in other words, is the image of desolation. Or, as Jean-Jérôme Carcopino remarked, it is "one of the handsomest bits of hell that one can imagine." Great flat stretches of black stone alternate with fields of needlelike projections upon which it is impossible to walk. But this black lava, uncomfortable as it was to touch, is beautiful to see. As it hardened, it took the form, in several places, of regular waves and graceful undulations. On top of these formations — some of them smooth, some rough — there were black blocks all around the camp, as though they had been carefully placed there by some gigantic hoist. And all this hard blackness frames a slender finger of golden sand.

How to Make Friends With a Sea Lion

When *Calypso*'s team had finished organizing their camp and had the time to sit down for a moment of rest, they realized that they were surrounded by animals. The nearest and boldest were the sea lions. Their fur touched with gold and shimmering in the light of the setting sun, they were making sea lion gestures — folding their necks, stretching out on the sand, and reaching their pointed muzzles and round heads toward the men and giving them looks which seemed more affectionate than curious.

Calypso's men had had long experience with sea lions. In the course of their expeditions, they had often had occasion to make friends with these animals. Two sea lions, Pepito and Cristóbal, captured off the Cape of Good Hope, had lived with us aboard *Calypso* for several months and dived with us, and had been free to come and go as they had wished. We therefore know that sea lions are highly individual creatures, that each one has a personality all its own and its own habits and whims. On the basis of this experience, it was possible for our men on Hood Island to live together in friendship with the sea lions whose home was the beach on which our camp had been established. They spent their days together; and, at night, the sea lions slept alongside *Calypso*'s men. The animals were not at all timid. They could be approached, petted — nothing seemed to upset them. They are, in fact, very tactile animals. The skin of the sea lion is unusually sensitive, and the way to establish communication with them is by touching them.

Probably the most salient trait of the sea lions of the Galápagos is their happy-go-lucky attitude and their *joie de vivre*. They played constantly — especially the females, who seemed to be extraordinarily friendly and curious.

Actually, the sea lions of Hood Island can afford to be carefree. There are so many fishes in the surrounding waters that food is no problem for them. They have no natural enemies other than sharks, and their agility in the water often enables them to escape even these great predators. Moreover, sharks are not very numerous in the Galápagos.

To appreciate a sea lion, one must see it in the water. They were at their best — and happiest — when they shared the divers' morning dip. The females were always the first to arrive and to rub against the divers. Then, when they were all playing, the divers would see an object like a torpedo streaking through the water toward them: the male, his fins tight against his body. He was bluffing, of course; but, even so, it was an impressive sight, for the male sea lion is quite large and has formidable teeth. Everyone would get out of his path; and the male, quite pleased at the effect he had produced, would continue on his way.

The sea lions' fondness for water games often generated scenes that everyone found very amusing. The male, who is a better swimmer than the female, even though he is less graceful and elegant, would sometimes be set upon by the females in a group. Often, one of them would grab him from behind with her flippers and allow herself to be towed along by him.

On land, the antics of the sea lions were equally comic. One could have sworn that the animals were lonely when the divers went out in the Zodiac for a shooting session. They would lie quietly on the beach, waiting for the men to come back. Then, when the Zodiac returned, they would waddle around excitedly with every sign of joyful agitation.

The sea lions quickly showed a particular affection for Jacques Renoir, who returned the sentiment wholeheartedly. The males, however — each of whom lorded over a harem of five or six females — did not always appreciate their wives' partiality for *Calypso*'s men. On land, they would growl, roar furiously, and pretend to charge. The charge, however, always stopped short of its target; and none of the males ever bit any of our men. The females did not get off so easily. Like any jealous lover, the male sea lion would then turn on his faithless mate and, in a raucous voice, let her know exactly what he thought of her behavior. Often the lecture was accompanied by a few blows with his head, or a sharp nip with his teeth. In such cases, the female would flee into the tent where she was safe from her indignant mate. Little Joe, the cook, was one of the most popular defenders of abused females; but if he was not available, there was always someone in the team to protect the animal from her mate.

Sea lions love to be petted — an endearing trait which, nonetheless, can become a bit annoying. When our divers and cameramen were working, for example, the sea lions were always in the way; and the film on which they were working after all, was not about sea lions, but iguanas.

Gentle Incarnations of Hell

There were a great many iguanas in the camp area, and *Calypso*'s men had more than sufficient opportunity to observe these reptiles with their dorsal crests, their dragon heads, and their staring eyes, like modern survivors from the Age of Reptiles. The opinions of the members of *Calypso*'s team had a wide range. Michel Deloire conceived a great admiration for the iguanas. He was especially struck by the harmony — even the physical harmony — which existed between these animals, with their spiked crests, and the island, with its lava spikes and wrinkles. Against this background, the iguana seems to embody the exotic, savage, and hostile nature of the volcanic isle.

Bernard Delemotte found something else to admire about the reptiles: "I liked the iguanas very much," he said. "Especially the red and green variety. They seemed to lead a freer life than the others, and to be more individualis-

On Hood Island, our divers and the sea lions live cheek by jowl in camp.

(Facing page) *Calypso* bids *au revoir*.

tic and more interested in what was going on around them. What I found most attractive about them was the way they handled themselves physically – their posture, the way they reared up their heads. When they come face to face with man, their appearance conveys a sort of attentiveness, but an attentiveness combined with pride. Sometimes they remain motionless for hours; and then they look like bronze statues covered with spots of enamel. That is when they are most attractive. They are less so when they move about, with little, hurried steps, stopping now and then to look around."

It is true that these iguanas have the appearance of monsters from the Secondary Era; but we must remember that they are only miniature monsters, and not the saurian giants of earth's reptilian age. The specimens found on Hood Island seem never to attain a length of more than forty inches. Those of Santa Cruz and San Cristóbal are slightly larger – almost four feet – but the variety found on Tower Island is a true pygmy and grows to a length of only about eighteen inches.

The iguanas look ferocious, but their ferocity is a matter of physiognomy and it is due above all to the shape of the head, the black eyes, and the claws. In fact, their appearance is more than a little diabolical, as though they were creatures who stepped out of a medieval monk's concept of hell. But appearances are deceiving. When we had occasion to pick up an iguana in our hands — which was surprisingly easy to do — we found that its skin was quite soft, and that its crest, despite its rough and jagged appearance, was not at all sharp. No part of the iguana's body is capable of inflicting a wound, or even of scratching. Moreover, they seem never to bite. They allow themselves to be picked up and handled with impunity, both on land and in the water. They may look like something conjured up out of the infernal regions, but they are actually the most accommodating of reptiles.

Despite this apparent willingness to acquiesce in anything that is done to them, it is extremely difficult to know what an iguana's reactions really are. One reason is that their eyes — like those of many other cold-blooded animals and of all species of fish — are totally expressionless. The eyes of the sea lions quickly reflect curiosity, or understanding, or affection. But a man can walk through a group of iguanas without a single one of them giving the slightest sign that it sees the intruder or even is aware of his presence. It is as though one is wandering among the gargoyles of Notre-Dame de Paris. But if we try to catch one of the animals, then we realize that it has a very wide range of vision, and one that extends far toward the rear.

The strangest physical peculiarity of the iguana is probably the appearance of its tongue. One would think that it would be long, slender, and sharp, like that of a lizard; but it looks more like a human tongue — red, rounded, rather thick and fleshy. Moreover, the tongue seems to play a major role in the behavior of the iguana. They use it often, for it is apparently the organ which serves as their sense of direction. They lick the ground, and then press the tongue against the palate on which the olfactory cells are located. It is in this bizarre manner that an iguana finds his way around.

These animals have another habit which is not less unusual. Our team quickly discovered that iguanas often spit on one another, and on any other animal — including man — which disturbs them. It is not really "spitting" in the ordinary sense, however. What they actually do is shoot a jet of water from their nostrils. This is made possible by the presence of a pair of saline glands, which are located under the skin, between the eyes and the nostrils on each side of the animal's head. There is a passage between these glands and the iguana's nostrils. The function of the glands is to filter out excessive salt from the sea water which the animal drinks. A considerable amount of this saline solution is accumulated in the glands, and it is frequently — and vio-

A male sea lion lodges a protest against his wives' unbecoming familiarity with our team.

lently — spewed out by the iguana. The iguana can "spit" about twelve inches; and when it senses danger, or when it fights, it uses this jet of water as a weapon of defense.

Origins Unknown

Someone once described the Galápagos as a laboratory without records. That is an accurate, if somewhat poetic, description. The reason for the lack of "records" is that the islands are quite young as islands go. That is, they are volcanic in origin and rose out of the sea at a relatively recent date. Therefore, there are no fossils in the Galápagos, and we do not know the ancestors of the marine iguana, or how or when they first came to these islands. The land iguana, on the other hand, is well known. It is found in South America, in the Caribbean area, and even in the Galápagos. But there are differences between the land iguana and the marine iguana — especially in the vertebrae.

The most obvious difference, however, is that only the marine iguana has a flat tail suitable for swimming.

The iguanas which we observed, in fact, seemed extremely well adapted to their aquatic way of life. They went into the water even when the sea was extremely rough; and they did not seem to be tossed about by waves which made it very difficult for our divers to work. Instead of going down directly to the bottom, where the water is more calm, the iguanas swim on the surface, among the waves, and with no apparent effort. The waves pass over them, but the animals continue moving away from shore. To return to land, they simply allow themselves to be carried in by the waves until they reach the rocks which line the shore, then they climb up, holding on with their strong claws. They never seem to lose their grip, no matter how slippery or smooth the rocks.

Since the iguana is able to handle itself so well in the water, one might conclude that the ancestors of the marine iguanas of the Galápagos simply swam over from the South American mainland. But that is far from certain. It is more likely that the crossing was made by iguanas clinging to pieces of wood adrift in the Humboldt Current. As to whether, at that time, the ancestors of the present-day marine iguana were able to swim, and above all to dive — that is another question, and one to which there is no answer available. In any case, no fossil of a marine iguana has ever been found in South America. It is possible that the iguana, having reached the Galápagos Islands and being unable to find food on land, took to the water in search of algae.

Why is it that the iguana, once it had learned to swim, did not cross the stretches of water which separate the various islands of the Galápagos? Why is it that, if indeed it crossed the five hundred or seven hundred miles of ocean separating the Galápagos from the mainland, it remained on the island where it had landed, and thus evolved differently from the iguanas which had landed on other islands of the archipelago? That is also a mystery. It may be that the iguanas were prevented from swimming among the islands by the strong currents that prevail there. Whatever the case, it is certain that the iguanas we observed never went farther out to sea than a thousand or twelve hundred feet from the island on which they lived.

Despite the chilling waters of the Humboldt Current, the sea around the Galápagos Islands cannot really be described as cold. The water temperature is usually comparable to that of the Mediterranean along the Riviera. For that matter, the ocean bottom in many respects resembles that of the Mediterranean, with its sea fans and its violet algae. When the water seems cold, it is only in comparison to the land temperature, which the sun, beating down on the rocks and the lava, turns into a veritable inferno. The difference

(Right) A blowhole on the south side of Point Suárez, Hood Island.

(Below) Little Joe, our American cook, comes to grips with the mockingbirds of the Galápagos.

between the land temperature and the water temperature is so great that to go into the water literally takes one's breath away. It probably has a similar effect upon the iguanas, whose body temperature is variable. They no doubt lose much body heat during their forays into the water, and much of their time on land is spent sunning themselves. Their sun-bathing habits, however, are quite different from ours. They take the sun in groups; and the groups are not only horizontal, but also vertical. One iguana will climb on top of another until there is a pile of them. Even in these somewhat provocative circumstances, however, it is rare that the iguanas will fight among themselves.

Usually, the individuals in a group of sun-bathing iguanas all face in the same direction. This is not an absolute rule, but it is certain that generally they arrange themselves in such a way as to achieve maximum exposure to the sun's rays and to take advantage of the heat reflected from the rocks on which they lie. So pronounced is the need for heat that, in order to obtain the greatest possible amount of it, an iguana will even adopt a vertical position to achieve maximum exposure to the sun.

The Chronicle of a Deserted Island

The following are passages from the diary kept by Bernard Delemotte during his stay at Hood Island:

February 8. At 6 P.M., *Calypso* blew its whistle in farewell as it rounded Cape Suarez.

After dinner, we all went together to find our new friends, the iguanas. We wanted to see where and how they slept. We found them stretched out on the black lava rocks, with their bodies flat against the rock as though to absorb the last calorie of heat from it. We walked through the iguana colony amid the most indescribable cacophony.

At 10:30 we returned to camp and found that we had to climb around the several dozen sea lions who had moved in and were sleeping there. They absolutely refused to move, and so we had to make ourselves as comfortable as we could among them. It was very hard to sleep. There were many baby sea lions, and they would wake up periodically and start crying for food, wanting to be nursed. The mother sea lions, like their human counterparts, would groan and mutter and pretend to be fast asleep.

February 9. Around 7 A.M., Michel Deloire and I went into the water with a 9mm camera and our battery-operated light. After only five minutes, however, we were back on the surface; the light did not work. We went back to camp, where Jacques Delcoutère and Henri Alliet set to work trying to repair

A marine iguana swims just beneath the surface.

François Dorado and an iguana come face to face beneath the surface.

the device. Michel and I then went out in the Zodiac to look around the area. In about six feet of water, we came across a group of iguanas. Apparently, these animals do not see very well in the water, and they did not react at all to our presence. We were able to observe them at our leisure.

Around noon, we returned to camp. After a quick lunch, we captured our first iguana and built a small pool for it. We have decided to keep three of them at the camp. So as to be able to handle them without hurting them, we have made harnesses for them with six-foot leashes. That should be enough for them to be able to run around a bit.

At 2:30 P.M., we went out again in the Zodiac, this time with our lights in working order. Michel and I dived in a calm spot, and then waited for the Zodiac to round up an iguana for us. Shortly, we saw one of the animals coming toward us at the surface. It dived down to the bottom, and soon we saw two small air bubbles escape from its nostrils. After a while, it raised its head, rose to the surface, and swam there for a few moments, with its head out of the water. Then, it dived to the bottom. We got some very good footage. Several times, we tried to use the 16mm camera; but each time, it jammed.

February 10. Problems. Michel Deloire wants to film the iguanas underwater, and Jacques Renoir wants to film them on land. I will have to divide myself in two. . . .

We have become aware of something unusual. Female iguanas, during the egg-laying season, take on extraordinarily bright colors which make them resemble males. The back of the head, the back, and a part of the arms are a luminous green. Their sides are variegated in red and black. The only enemy of the iguanas that we have been able to discover is a variety of thrush, which eats the iguana eggs. Strangely enough, the female makes no attempt to chase away these predatory birds.

The individual male iguanas live with their females in a well-defined territory. If a strange male intrudes, there is a battle which lasts until one of the males is decisively beaten. The males do not seem to fight to the death.

The "territory" of an iguana is usually a crevice in the rocks in which sand has gathered. The female lays her eggs there and buries them. Even when the mother must temporarily leave this nest for some reason, she keeps looking back at it and returning to inspect it, for such places are hard to find and it happens regularly that another female comes and tries to lay her eggs in the same place. Then, the two females fight — and the issue is usually resolved by the male, who makes peace by hitting the females with his head.

The Darwin Foundation experts have assured us that the iguanas of Hood Island are different in color and behavior from those of the other islands. Each island, in fact, is a separate world; and it is this, as Darwin recognized, that makes the Galápagos such an ideal laboratory for studying the laws of behavior, ecology, and evolutionary adaptation to the environment.

There is a large rock protruding from the water south of Point Suarez, and it is covered with iguanas. It seemed to us that, since there were so many animals on the rock, there must be some also in the water around the rock. This seemed a good opportunity to observe the animals feeding — but we were disappointed. It may be because there was such a strong swell. We were working in about ten feet of water, and we were constantly drifting into the troughs between waves. Conditions were so bad that we were unable to get a single foot of film. And, because of the constant changes in depth, we all had earaches.

About six o'clock in the evening, we saw a herd of goats walking along the bay. Little Joe pointed out that we need fresh meat — and Dr. Perry has advised us to kill as many goats we we can so as to preserve the rare fauna of the islands. I went out immediately and shot a young male. I also found two baby goats which had apparently been abandoned by their mother. Joe has adopted them and found a small cave in the rock where they will be protected from the hot sun. As for the goat which I shot, it was quickly cut into pieces, salted, and then buried to keep the meat cool.

Not all of Little Joe's problems are so easy to solve. He is constantly plagued — by mockingbirds. As soon as Joe begins preparing a meal, the birds cluster around the table, peck at the food, and strut around on the plates and in the pots. Joe yells and waves his arms, but the birds ignore him completely. We have never seen birds as brazen and as fearless as the mockingbirds of the Galápagos.

February 11. It is generally believed that iguanas prefer places where the sea is calm, so that they can get in and out of the water without difficulty. Our observations, however, indicate that whether the water is rough or calm is a matter of indifference. I have often seen iguanas in very rough water, and they simply float among the waves without appearing to be bothered in the least by them.

We spent the entire morning today trying to find iguanas in the act of feeding on the bottom; and we were unsuccessful. In the course of our search, we inspected hundreds of yards of cliffs. We explored a dozen caves, some of them with two exits. All this in water from 30 to 60 feet deep.

During the entire dive, we were accompanied by eight or ten sea lions who spent their time in playing the most imaginative games. One of their favorites — and one of ours — was to imitate everything that we did. In the caves, for example, we would stop and look back toward the entrance; and

Michel Deloire and Yves Omer lie in wait to photograph an iguana on the beach.

the sea lions would stop and look back toward the entrance. Also, they were apparently jealous of the quantity of air bubbles released by our diving equipment, and they expended much effort and air in trying to release an equal quantity.

The interior walls of these caves are fantastically beautiful — covered with sea fans, sponges of all kinds and shapes, and sea squirts. In one cave we found two trees of black coral. One had branches of bright yellow; the other's branches were green and tufted. The trunks of both were quite large — about as big around as a man's fist.

The beauty we enountered during our dive, and the antics of the sea lions, cannot make us forget the fact that we have failed so far to film a single iguana in the act of feeding.

Marine iguanas are good swimmers and are relatively easy to approach in the water.

Jacques Renoir, meanwhile, has had better luck. He was able to film iguanas feeding among the rocks at low tide, against the background of a waterspout shooting from 50 to 75 feet into the air. This spout is located on the south side of Point Suarez; and, since the prevailing wind is from the south, the spray reaches our camp and we spend most of our time in damp clothes.

February 12. Today the sky was cloudy and, in addition to the usual southeast swell, there was a heavy swell from the west. The waves were so high, and the water so rough, that we could not work in our usual spot, and so we set out to find another location along the southern coast of the island. The first small bay was too rough, and so we went on to a larger bay a short distance away. There we found a rocky isle covered with iguanas. There was a six-foot swell, but the isle was high enough to be protected from it.We therefore decided to dive there, and we were pleasantly surprised to find the water quite clear. The rock walls of the iguanas' isle were covered with sea fans, sponges, and algae of violet, green, and ocher moving in the swell. From beneath the surface, we could see that the base of the isle was shaped like an arch, with an opening in the middle; and we allowed ourselves to be carried through this opening by the swell. On the other side, we saw a field of rocks covered with algae — a prairie of algae in water from 10 to 50 feet deep.

Suddenly, we stopped. We could hardly believe our luck. Directly in front of us, an iguana was swimming nonchalantly toward the algae. It lighted on one of the rocks and began eating. Instantly, Michel Deloire and I went to work. Michel, trying to keep his balance at a ridiculous angle in the swell, began shooting, while I handled our battery-operated lights. But we had no sooner begun than the iguana swam away to another rock. The light from the projector caught the animal; but it was frightened and swam away again. This time, we lost him. But we soon got over our disappointment. Another iguana came down, and then another. We could hardly believe our eyes. We had been trying for days to shoot this scene.

Iguanas are not very good swimmers, but they are remarkable divers. They are able to go down very rapidly; and they move quickly once they are on the bottom. In fact, they are able to outdistance a diver; and so, once we lose sight of them in the water, there is very little chance of finding them again.

In this instance, we decided on a special technique to enable us to resist the swell. I took a firm grip of a rock and then grabbed Michel by the belt. In these waters, apparently, the last thing a diver wants is mobility. What he needs is muscle power. At 12:45, we climbed aboard the Zodiac, exhausted but happy. We had shot 350 feet of 16mm film, and 300 feet of 9mm.

There are no large kelpweeds in the Galápagos. The food of the iguanas is a reddish and rather short algae which grows on many of the rocks in these waters. This algae resembles a small sea fan. We have tasted it, and found it rather hard and crisp — an indication that it is rich in limestone.

February 14. The first thing this morning, we took a detailed inventory of our equipment and supplies. We are very low on both food and water.

There was still a heavy swell, and the tide was not yet completely in. There was a strong undertow, and we had to wait. At eleven o'clock, we saw an iguana slip into the water. That seemed as good a moment as any to begin.

The water was particularly cold, and the currents in the rough water were a constant hindrance. After a long wait, we saw an iguana swimming on the surface. We began moving slowly toward the animal — but apparently not slowly enough. He dived — which astonished us, because the water was 45 or 50 feet deep at that spot. The depth, however, did not seem to bother the animal. It swam gracefully toward the bottom in a zigzag pattern and, after standing on a rock for a few moments, began feeding on turf of violet algae. For at least fifteen minutes, we were able to film the animal uninterruptedly, shooting with a suited diver in the foreground. At the end of that time, the iguana calmly swam up to the surface, with algae floating from the corners of its mouth. Then it dived again for more food, and we were able to continue filming.

We noticed that the iguana does not remain still for a single second in the water. It eats constantly; and it eats only certain algae which it likes. It uses its cutting teeth — which end in three points — to tear these from the rock. It was the first time that a marine iguana had ever been filmed while feeding in the open water.

We introduced a few human elements into the sequence. At one point, for example, the film shows Jacques Delcoutère giving a bit of algae to the iguana, and the iguana taking it. This was of particular interest to us, because it is generally believed that the marine iguana never accepts food from man, and that, in captivity, it inevitably starves to death. But here, in the open water, where there was a great deal of algae available, we succeeded — by being very very careful — in hand-feeding the animal its favorite algae.

I wonder if we will ever be able to tame an iguana. In spite of our success today, I am inclined to doubt it. So far, no iguana has shown the slightest sign of recognizing us — or even of seeing us. We, at least, have reached the point where we can begin to tell one animal from the other by their size, or a particular marking, or by a scar — probably the consequence of an attack by a shark.

In the evening, after dinner, Jacques Renoir began filming some background shots. Michel Deloire and Henri Alliet fiddled with their underwater camera. Jacques Delcoutère wrote a letter. François Dorado plucked his guitar. And we could hear Joe banging around in the kitchen, and the goats calling. I wrote in my diary for a while, and then I walked around the camp. Everything was peaceful. The baby goats were in their cave; and the iguanas were stretched out on their rocks, absorbing the heat remaining from the blazing sun.

February 18. At 4 P.M., Joe sees a herd of goats through the binoculars. Immediately, Michel Deloire takes off, armed to the teeth. He returns an hour and a half later with two shoulders and two legs. He swears that this is the first time that he has ever killed an animal, and adds, with a sigh, that "we must really need meat for me to do something like this." He is right. Little

Marine iguanas and sea lions live together in harmony—despite the sea lions' fondness for teasing the reptiles. They delight especially in pulling the iguanas' tails.

Joe's supplies by now consist of cans of fruit. He has an especially abundant store of pineapple slices in heavy syrup, the very thought of which is enough to nauseate us.

During the afternoon, we shoot a sequence of sea lions swimming on the surface and pulling the iguanas' tails. A very good scene. The sea lions regard the whole thing as a game; and the iguanas, although they obviously do not see the fun in it, seem more distressed than frightened.

Late in the day, we load a few supplies into a Zodiac and wave good-by

to Jacques Renoir as he heads out into the bay. He is going to Fernandina Island, to Cape Espinosa, to film the egg-laying of the iguanas.

February 20. When we returned for our morning's filming at half past noon today, we were greeted enthusiastically by our colony of sea lions. They seemed to be in the mood for games, and Michel took advantage of their playfulness to film them underwater as they repeated their performance of yesterday and amused themselves by teasing the iguanas and giving them little nips.

The rules of the game are unvarying. The sea lions begin by finding an iguana sunning itself on a rock in the bay. They prod it and poke it until the reptile decides to get away from the intruders. Once it is in the water, the iguana swims quickly away, trying to outdistance the sea lions. It is a real race; but one which the sea lions always win, for they are much faster than the iguana. When the sea lions have caught up with their victim, they play with it as a cat plays with a mouse; the only difference is that the iguana is never hurt. Only its dignity is wounded.

February 22. The weather is very bad today. The water is cloudy, the swell is enormous, and the sky is heavily overcast. It was impossible to dive, so we decided to explore the base of the cliff. To our great surprise, only 500 feet from camp we find an iguana digging its nest. In a little over a mile, we go from surprise to surprise. There are a large number of iguanas inland. We concluded that the season for laying eggs has begun. Old nests were being cleared out and made ready for occupancy, and some new ones had been completed. We found about forty iguanas busily digging others, moving the small amount of sand and soil that had accumulated among the rocks.

February 23. Today, fortunately, the sea is not quite so rough, and we went out to the iguanas' feeding grounds for some shots of the animals both feeding and climbing out of the water. It was very difficult for us to get the sea lions to leave us in peace long enough to get some work done. I was foolish enough, the first day we were here, to let some of them go for a swim with me, and now we cannot get rid of them. Just as we were about to give up, however, an old male showed up and herded the females back toward the beach.

February 25. Early this morning we set out to get water from a pond that we had been told about. Twice on the way we had to stop and change our Zodiac's sparkplugs. Finally, we landed on a magnificent white beach and, loaded down like workhorses with containers, we headed inland. At first, we had to use machetes to cut our way through the thickets; but, after a few yards, walking became easier. In some places, the terrain resembled the African bush country. It took about three quarters of an hour to travel the mile to the pond. There are actually two ponds, each one about 800 yards across — and they were both completely dry. All we could see were rocks and red earth, the latter riddled with cracks two feet deep. There seemed to be little fauna in the area — a few finches, mockingbirds, and lizards. We saw no land iguanas.

On the way back to the beach, we stopped for a while on the north face of a hill about 300 yards high. It appeared to be made up of red earth and fine gravel, and there were many holes caused by erosion. The island's goats, who live in such holes, have a plentiful supply of ready-made homes.

We reached the camp at 1:30 P.M. At 4:00, we sighted *Calypso* in the bay.

CHAPTER THREE

The Enchanted Isles

Calypso, having taken on supplies of fuel, water, and food at Guayaquil, began the return trip to Hood Island. The wind was moderately strong, the weather rather bad. Our first night out, however, we were able to establish radio contact with our camp at Point Suarez, and it was reassuring to know that everything was going well.

The weather improved during the following day, even though we were still rolling quite a bit. In the latter part of the afternoon, we dropped anchor off Point Suarez, and the team we had left ashore there immediately came aboard. They all seemed to be in excellent shape. Deeply tanned, almost naked, they looked like modern Robinson Crusoes who had opted for the life of primitives, far from civilization. When we told them of some of our difficulties at Guayaquil, they seemed to think that we had been silly to worry about such trifles. Yet, the black lava rock on which they had been living, broken only by an occasional patch of bushes or euphorbias, looked singularly unattractive to us, who had had the advantage of *Calypso*'s modest comforts. Despite the state of pure nature with which they professed to be so content, they seemed happy to receive the fresh supplies which we had brought from Guayaquil.

We spent the following day at anchor in the bay, which allowed us to undertake several missions simultaneously. Ron Church took underwater photographs, with the aid of a diver and a launch. And Yves Omer undertook an exploration of the bottom around Point Suarez. Close to *Calypso*, he discovered an exotic fish which, thanks to material provided by the Darwin Foundation, we were able to identify as a batfish (Ogocephalidae). In the spirit of Darwin, we were curious to know how evolution had arrived at the batfish's particular form, and how the singular features with which it was endowed by nature contributed to the continuation of the species. A batfish is not large — about the size of a human hand — and lives on the bottom in depths of between forty-five to sixty feet. Its appearance is unusually striking. Its mouth, with its red lips and downturned corners, resembles that of a sad clown. On top of its sinister head, there is a horn, which ends in a white growth; and this serves as bait to attract the batfish's prey. Its pectoral fins resemble paws; and it is likely that it supports the forward part of its body on these paws, while the rear part rests on its tail.

The batfish remains absolutely immobile on the bottom, but its reactions when one tries to capture it are unusual. As Yves's hand reached down to grasp it, the batfish waved its tail and took a leap forward, using its pectoral fins, or paws, like landing gear to settle on the bottom again. This, apparently, is its chief method of locomotion, for the batfish is a very poor swimmer.

Bernard Delemotte, François Dorado, and Henri Alliet set out to film the batfish and succeeded in capturing several specimens in a plastic hemisphere. We wished to keep them aboard *Calypso* so as to be able to observe them at our leisure. Any scruples we had about capturing specimens were set at rest by the fact that batfish, as strange as they appear to us, are quite common in these waters. This area, in fact, contains an extraordinary number of all kinds of exotic and colorful fishes. There are five species which are found nowhere else in the world.

The batfish aboard *Calypso* were kept in a special and highly specialized aquarium designed by Commandant Alinat. Its upper third consists of a "roof" divided into several sections which fit exactly into the bottom two-thirds, which is the aquarium proper. All the pumping, aerating, and filtering systems are contained in this roof; and everything works on separate circuits. The particular advantage of Commandant Alinat's aquarium is that it is proof against rolling and pitching. This is very important to the comfort of the aquarium's inhabitants; for fish, paradoxically, suffer from seasickness. We have discovered this through long experience; for we have had several specimens die aboard *Calypso* simply because they were kept in ordinary aquariums.

In our special aquarium the captured batfish were quite comfortable. Their appetites were so healthy that they ate everything within reach of their mouths, even though they are unable to turn their heads. We never tired of watching their "take-offs" and "landings."

Iguanas have a ferocious look, spiky crests, and enormous claws. Yet, for all of that, they are inoffensive creatures.

On Another Planet

One of the questions we asked the team from the camp, once they came aboard, was how they had felt about their isolation on Hood Island. They thought the question was irrelevant, for they had adapted quite easily to the rugged life of the island. Also, they had had enough work to keep them busy from morning to night, and not much time to brood over the fact that they were on a deserted island, in the middle of the Pacific, entirely on their own.

Bernard Delemotte expressed the feelings of his friends very well. "We felt quite at ease on that island," he said. "After all, we had interesting surroundings, and we were in the midst of a colony of fantastic animals. The most rewarding thing about it, I think, was that it *was* so strange; it was as though we were living on another planet."

We wondered what the men and the iguanas had thought of one another. The men from the camp explained that they had ended by developing a real sympathy for the reptiles. But, speaking for the iguanas, they admitted that, as men, they had met with complete indifference on the part of these animals. The iguanas showed not the slightest hostility or antipathy for the strange beings who had disturbed them, handled them, and even dragged them down to depths to which they were not accustomed. But neither had they, like the sea lions, come to recognize individual divers or become attached to any of them. The over-all impression conveyed was that the iguanas wanted only to be left alone to eat algae and sun-bathe.

It is undeniable that this first camp on Hood Island played an important part in the morale and spirit of our team. Our divers and cameramen, left alone on one of the enchanted islands of the Galápagos, had been able to take the measure of their environment and of its exceptional aspects. Living as they did in the midst of those living fossils, the marine iguanas, and establishing an amicable relationship with the sea lions on the beach, it was as though they had discovered a world forgotten by time and change. In their careers on the sea, they had become accustomed to moving from place to place. Then, suddenly, they had been left to shift for themselves on a desolate, isolated bit of lava rock with animals who had never seen man, and therefore never known fear. It was like being alive when the earth was young. And not the least important result of their experience was that this first "adventure of *Calypso*" in the Galápagos awakened in the whole team an intense curiosity about everything in the archipelago that we were about to explore.

On February 27, we dismantled the camp on Hood Island; and, in the afternoon, *Calypso* headed for Fernandina Island, where we were going to establish a second camp at Point Espinosa. We had already reconnoitered

The batfish (*Ogocephalidae*), a marine oddity, abounds in the waters of the Galápagos.

Point Espinosa, and its volcanic formations, its flora, and its fauna all seemed to differ from what our team had known on Point Suarez. Even the iguanas were different. Instead of being red and green, on Fernandina they were dark gray.

In the Galápagos, one can be certain of finding either a species or a variety of animal peculiar to any given island. Each of the islands, no matter how small it may be, seems to be a closed universe in which animals evolve, perhaps for thousands of years, without outside influence or interbreeding. This is what specially struck Darwin. It is certain that there are very good reasons why the animals remain in a particular region, either because the conditions in that spot correspond to the animal's needs (the iguana, for example, finds the kind of algae that it likes), or because the animal has adapted itself to certain kinds of food or flora — which is the case with the finches.

In the morning, Bernard Delemotte helped the team for the new camp to select a place and to set themselves up there. The island is small, and there are not many good sites. There are only three men in the camp: Jacques Renoir, our cameraman; Jean-Jérôme Carcopino, his assistant; and Jean-Clair Riant, a diver.

Bernard then explored the entire area between Point Espinosa and Point Douglas. The bottom seemed interesting; but everywhere in the archipelago there is an enormous swell which breaks against the rocks from south–southwest.

One of the first things we noticed was that the lava formations on Fernandina are different from those on Hood Island; they are more rounded and less sharp. Also, Fernandina is richer in vegetation and, at first glance, in

animal life. In addition to the gray iguanas, there are many birds: boobies (or gannets), penguins, and herons. Especially plentiful are the wingless cormorants which, despite their name, have truncated wings but are nonetheless unable to fly; instead, they swim and dive and feed on fish.

Jean-Paul Bassaget and Ron Church, while the camp was being set up, decided to indulge in a bit of mountain climbing. The goal was the crater which dominates the island. The project was more difficult than it had seemed. First, there was a mass of lava to be scaled — and this lava was not rounded but covered with sharp, cutting edges upon which it was extremely uncomfortable to walk. There was absolutely no path; and in the extreme humidity and heat, what had begun as a lark became an ordeal. The view from the top of the volcano, however, made the effort seem almost worth while. Jean-Paul and Ron could see the entire island and the surrounding islands. The crater, they reported, was a blue lake; and in the middle of it was a smaller crater containing a green lake. Fernandina itself, seen from above, struck one immediately for what it was in fact — the result of a volcanic upheaval, covered with black and red, and relieved only by the green of the trees and shrubs.

The coasts of Fernandina Island are steep black cliffs which end in the sea. These cliffs are not very high, but they have a forbidding look about them because of the deep slashes in their bases — against which the waves pound with a noise like thunder. Yet, one has only to round one of the island's capes to discover a stretch of water as calm as a lagoon. And this is typical of the

(Left) Jean-Paul Bassaget, standing at the highest point of Fernandina Island, inspects the island's crater and lake.

(Right) A land iguana sees its reflection in a mirror; but, unlike the grouper, it shows no aggressive reaction.

(Below) A group of marine iguanas looks up in curiosity as members of our team approach.

islands of the Galápagos. There is one steep and inaccessible side, and one which slopes gently into the sea. The cliffs are on the west or south sides, while the slope is on the north.

At Point Espinosa, the lava has formed a sort of amphitheater above the level of the sea, with circular tiers — and this is the favorite spot of the penguins and of a colony of small sea lions which, since they have never been hunted, are perfectly at ease with human beings. It was this amphitheater which Bernard Delemotte and his friends, after having circled the island in a Zodiac, selected as the most suitable site for our second camp.

Fernandina Island has a number of small shoreline canals in which the water is always calm. These are separated and protected from the open sea by walls of lava rock. The canals are the home of sea lions who have offspring;

and there one can observe the calves nursing. Mangroves have grown around these lakes; and, surrounded as they are by black lava, the water and trees seem an oasis of coolness and tranquility. The iguanas are apparently of the same opinion, for some of them live among the mangrove roots.

The establishment of our camp was regarded with indifference by the iguanas of Fernandina. It was not difficult for us to pretend that we, for our part, were also oblivious to their presence; for, lacking the bright colors of the iguanas of Hood Island, they are difficult to see against the black lava. It is possible to pass close to a group of these reptiles without being aware of them.

Here, as on all of the islands of the archipelago, goats are numerous. The men who were about to move into the new camp gave us notice beforehand that they had every intention of adding variety to their meals by butchering a few of them — as the director of the Darwin Foundation had urged us to do.

These goats are quite wild; and, in addition, they are so agile that it is difficult to catch them. It is interesting to note that, in the Galápagos, none of the wild animals native to the islands have any fear of man, while the animals which were once domesticated and then abandoned by mariners are extremely wary of humans. This is true not only of the goats, but of the cattle, horses, and dogs which have reverted to the wild state.

Unexpected Enemies

It required an entire day for us to set up the camp. There was material to transport to the beach, and the camp itself had to be erected. Finally, at 9 P.M., *Calypso* was able to raise anchor and sail out of the bay toward the Island of Rabida. On Fernandina, we left our three new Crusoes.

The men in the camp were not at all apprehensive. After all, they had a camp comfortably situated among the mangrove trees, in a location which was both dry and, at the same time, provided easy access to the sea. Their first evening on the island, however, they discovered that they had enemies on Fernandina. *Calypso* had hardly disappeared from sight when the camp was invaded — by mice. These little rodents were gray, and quite handsome; but they had been endowed by nature with an audacity, and an appetite, before which nothing edible was safe. It was a great crisis, and *Calypso*'s men thought themselves equal to it. Some of their provisions, they suspended by ropes from the branches of the mangroves; and others they buried in holes in the ground. And then they settled down to sleep in the knowledge that, after all, men were smarter than mice.

The mice, apparently, thought otherwise. And, throughout the night, they worked to prove their superiority. One of the men awoke to find the mice building a living pyramid of their own bodies in order to reach the provisions hanging from the mangrove trees. And Jacques Renoir, who was sleeping in the open air in a hammock, was rudely awakened by a particularly audacious mouse who was nibbling on the most sensitive part of our cameraman's anatomy.

Fernandina is the only island on which we encountered these pests. Their ancestors, no doubt, jumped ship here at some time in the past.

The next morning, another enemy showed up. But this time it was a familiar one: the mockingbirds. Jean-Jérôme Carcopino expressed concisely the irritation of his teammates at these new enemies. "It is tempting to succumb to their charm," he said. "They are very cute, and they seem to be very curious about everything that you do. The trouble is that they never leave you alone. They follow you everywhere; and soon you begin to get a little tired of them. There are so many of them that there is no place to go to get away. And when they begin pecking at the food on your plate and drinking out of your glass; when there are forty of them on your table and nothing can make them move — at that point these adorable little birds can drive you straight up the wall. And, of course, we didn't want to harm them, let alone kill them, as appealing as the thought might be. There was nothing to do but pick up our plates and glasses and go to eat in the tent, with all the flaps closed."

The Mating Season

In the morning, the iguanas, which we had seen the night before packed together in groups, were spread out. Their dark color and shining bodies made the landscape seem even more strange than it had the preceding day. The men in camp were especially aware of the harmony which seemed to exist between their surroundings and these dark dragons, covered with spikes, resembling monsters risen from an extinct crater, as though they were themselves creatures of lava. Even the immobility of the iguanas was in accord with the black petrification of the island.

It was an alien environment; but not a hostile one — or at least not one which our team regarded as hostile. To the contrary, the men found it strangely pleasing, and even calming. There was a strange sense of peace and relaxation. To outward appearances, Fernandina seemed inhospitable to life forms. And yet, life was abundant there, and the animals had succeeded in adapting perfectly to conditions which were apparently impossible.

At Point Espinosa on Fernandina Island, the shore is literally covered with marine iguanas.

Our team's mission on Fernandina was to film the mating of the iguanas; to observe the nest-building of the females; and, if possible, to observe the laying of the eggs. The first attempts, however, were disappointing in their results. The iguanas were so secretive about mating that the team came to believe that they mated only at night. During the day, all that could be seen were iguanas in piles, motionless in the sun. Occasionally, one of them would move toward the water for feeding. A few more would follow. But there was no activity connected with mating. There was no way for our men to know that the laying of eggs had already begun.

Puzzled by this apparent lack of activity during the day, Jacques Renoir and his companions decided to surprise the iguanas during the night and film their mating. In the middle of the night, therefore, loaded down with flood-lights and cameras, they crept toward the sleeping animals. Suddenly, they switched on the floodlights. There was instant pandemonium. The iguanas had been sleeping in piles in even greater numbers than during the day. There must have been at least five or six thousand of them in the lava amphitheater that night. Awakened by the lights, they panicked and tried to escape, crawling over one another and even trying to climb up the legs of our men, inflicting deep scratches with their claws.

Meanwhile, the sea lions, sleeping nearby, had been awakened by the din and they too became frightened. The females tried to escape from the imagined danger and became hopelessly embroiled with the iguanas; while the males, roaring, charged back and forth aimlessly.

The whole thing was a nightmare. The camera team switched off the lights and went disconsolately back to the tent. Little by little, the iguanas and sea lions quieted down. Finally, everyone slept.

The Battle of the Nests

It did not take our team very long, after their nocturnal fiasco, to discover the secret of the iguanas' mating. They noticed that the females, unable to find suitable places for nesting along the beach, were laying their eggs at the tops of the cliffs, some 150 to 175 feet above the beach. It was only there, it seems, that they could find the crevices and the loose earth and sand which they needed to dig their nests. Jacques Renoir, in exploring these heights, found nests from the preceding year, and even the remains of broken eggs. The most extraordinary thing about this was that the females must have been capable of performing feats of acrobatics in order to reach the nests, for the cliffs are almost vertical. The highest and most difficult point is a large rock — which our team called "the castle" — and there were many nests even there. There was nothing for our cameramen to do but follow the iguanas up those walls of lava, carrying all their equipment.

The female iguanas, once they reach their nesting place, often fight among themselves. As soon as one has dug her hole, another one shows up; and, being too lazy to dig out a nest of her own, she tries to take possession of the other iguana's nest. The battle, however, is never very savage, and intimidation plays a large part in it. The two females push one another, and their heads clash; but they use neither their teeth nor their claws.

Our team was able to film the arrival of a male during one of their encounters. It promptly separated the females, and apparently ordered the lazy female to go dig a nest of her own. In any event, the first female was left in victorious possession of the nest she had dug.

After spending several days in observing the iguanas, the men noticed that the animals did a good deal of fighting among themselves. The females fought for nests; and the males fought for the females. The males seemed particularly bellicose. They were always ready to join any fight in progress, no doubt as a means of asserting their authority. But their idea of joining in was usually to raise their heads threateningly, and to balance themselves on their forepaws, rather than to attack. Even so, occasionally there were really fights; and it happened sometimes that one or another of the animals was wounded. Even in such cases, however, the adversaries do not use their teeth or claws, but resort to butting their skulls together.

The nests dug by the females are tunnel-shaped, and are usually twelve to sixteen inches deep. The females slide their eggs into this tunnel. The eggs are soft and have no shells. Hatching takes place after sixty days — if the eggs have not been eaten by snakes, which are the greatest danger to the perpetuation of the species.

The female guards her eggs while they are incubating; but she never sees them, for she never looks into the tunnel. Instead, she puts her tail into the tunnel, and thus is assured, by purely tactile means, that her eggs are safe. If one of the eggs is replaced by another object — a golf ball, for example — the mother iguana does not seem to notice the difference. But if one egg is taken and not replaced by a similar object, she is obviously distressed and does a great deal of spitting and hissing.

Territory

It is not known for certain whether iguanas have individual "territory" or, if they have, whether they have the will to defend that territory. They are gregarious animals and live cheek by jowl, so to speak, in tight groups. They do not seem to fight in defense of a territory in the proper sense of that term. Their encounters, according to our observations, take place only for the possession of a particular nest — a rather different thing. The reason for these conflicts is the relatively small number of suitable sites for nesting. On the other hand, in order to reach their nesting places, the iguanas are often obliged to cross the territory of the boobies — birds which have a very keen sense of what is theirs. Whenever an iguana — or anything else — crosses the space which the boobies regard as their property, it runs the risk of being attacked. The iguanas seem aware of this and try to be on their guard; but occasionally they are caught unawares.

We ran the same experiment with the iguanas as we had with several other marine animals, that is, we placed mirrors in front of them. The results obtained were somewhat different from those in previous experiments. We had discovered that groupers, when presented with their own reflection, think they are in the presence of a rival—and attack the mirror. The common octopus (*Octopus vulgaris*) had manifested curiosity and rubbed the mirror at length with its arms. The iguana, however, did nothing except watch its reflection. Even if it thought the reflection to be another iguana, it showed no sign of an aggressive reflex. This was not an unexpected reaction, for the iguana is a social animal.

Other than boobies and snakes, the iguanas' most constant enemies are wild dogs and sharks. We frequently saw iguanas without tails — obviously the victims of encounters with one or the other of these enemies. Some birds, however, are friendly, or at least useful, to the iguanas. Finches (*Geospiza difficilis* and *Geospiza fuliginosa*), for example, pick off the ticks which attach themselves to the iguanas' skin.

Perhaps the most salient characteristic of the life-style of the marine iguana is its inactivity. Piled up one against another, the reptiles remain motionless for hours on end. Carcopino used to say that it was as though a film had suddenly been stopped at a particular frame. They would all be moving slightly — then in an instant, the entire group would freeze and remain immobile for a certain length of time. Then, one iguana would move, and the rest would follow suit. Next, another period of immobility. And so forth. The life of an iguana is composed of starts and stops.

These reptiles spend many hours doing nothing but warming themselves in the sun on the rocks. They go into the sea only when the tide is out, in order to feed on algae in shallow water. Feeding, however, is not a group activity. Individuals go alone and return when they have eaten their fill.

So far as we have been able to determine, there is little communication or contact, other than physical proximity, among iguanas. Which leads us to the usual question: "Are iguanas intelligent or stupid?" The best answer is that given by Yves Omer: "I have never met a really stupid animal." It must be admitted, however, that iguanas do not seem gifted with particularly lively reactions, and that their psychological equipment, like that of many reptiles, is weak.

One of our purposes in establishing a camp on Fernandina Island, as I mentioned, was to film the mating of iguanas. Iguanas, however, are very shy about their mating habits; and although we sometimes surprised animals in what appeared to be compromising positions, we could never be certain that they were actually mating.

When iguana eggs hatch, there are as many young males as there are females. By the time adulthood is reached, however, the males are much less numerous than the females. We have often wondered why. After all, fights among males are never so deadly as to have an effect upon the iguana pop-

Marine iguanas dig their nests in the sand and defend them against intruders.

ulation. The most reasonable explanation seems to me to be the following: The newborn iguanas all feed, at first, in puddles along the beach, where there is no danger. As they grow larger, the males venture into the sea to a depth of perhaps twenty-five or thirty feet. At that depth, they encounter sharks. It is not known why young males dive deeper than females; but it is certain that all the iguanas that our divers met in relatively deep water were males.

Our team noticed that marine iguanas, when frightened, never take refuge in the water; instead, they try to escape inland. This may be an indication that the species originated on land rather than in the water.

On Fernandina Island, there is a certain species of crab (*Grapsus grapsus*), which is red with a blue underside. This crab plays an important role: it is the gravedigger and the beach-cleaner of the islands, for it feeds on the cadavers of dead iguanas. Sometimes there was a large number of these crabs on the shore, but they were never seen to go into the water.

Aboard Calypso

Since we wanted to use to greatest advantage the time spent in the Galápagos, we had three projects under way simultaneously. While Jacques Renoir and his companions were on Fernandina Island filming the egg-laying of the iguanas, we had men on Hood Island filming iguanas in the water. At

Returning from a dip in the ocean, the iguanas must cross territory claimed by the "boobies." The birds are quick to let them know what they think of trespassers.

the same time, *Calypso* was on a tour of the islands. Occasionally, we dropped anchor at one or the other of them, such as Rabida Island, from which we could see the other islands of the archipelago stretched out before us in a circle. Some of these islands had a reddish appearance; others were covered

On the black rocks of the Galápagos Islands, a crowd of red crabs (their undersides are blue) functions as ditchdiggers.

with vegetation; and others were completely black. It was a fascinating diversity of landscapes — particularly since we were aware that there was a corresponding diversity of animals and ways of life.

On Plaza Island, we came across land iguanas. These reptiles have a true golden color. We have only rarely seen animals of that hue. They seem to be wearing armor of gleaming gold.

Land iguanas feed on leaves and flowers, and they eat even opuntia leaves — which have spines as sharp as needles. These iguanas have suffered much from wild dogs; but since the Galápagos have become a tourist attraction, they have found a new source of food. They beg food from visitors, and they are extremely daring about it. Their favorite dish is bananas. Unlike marine iguanas, they do not "spit." But they have been known to bite.

Floreana Island is inhabited by flamingos. It was near this island that Bernard Delemotte and Michel Deloire came face to face with a large hammerhead shark while they were filming turtles in a strong current. The shark came to an abrupt halt about twelve feet from our men, and then turned one side of its head toward them. They could see its eye, at the end of the "hammer" closest to them, staring at them attentively. It was a frightening moment. They had no idea whether it would attack. But, after having inspected Bernard and Michel, the shark lost interest and swam away.

As often as we have encountered hammerhead sharks, it has always been in water so foamy that it was impossible to see them clearly. These animals seem to have a preference for areas in which the waves break into foam. It is said that the hammerhead does not see very well; but we know from our own experience that these sharks make a particular effort to take a close look at divers. To do this, they turn their head to one side so that they may have a good view from one of their eyes.

The island of Genovesa looks like a crater the top of which has been broken off. Its walls are almost vertical. Within the crater there is a large cirque, at the bottom of which is a blue lake. Genovesa is the home of a large number of frigate birds. We were able to photograph them, along with red-footed gannets and blue-footed gannets.

One of the most beautiful of the Galápagos Islands is Gordon, which is shaped like a sugar loaf. The base of the island, beneath the surface of the sea, has many holes — the work of sea urchins. We dived often, for the water is crystal clear. Michel Deloire did a good deal of filming, some of it at depths up to two hundred feet.

The only airport of the archipelago is on Baltra Island. I landed there when I arrived in the Galápagos to join the expedition and to study the results of the work already performed by our teams.

Between Baltra and Santa Cruz there is a shallow, narrow canal. There were times when *Calypso*, in navigating this canal, had barely three feet of water beneath her keel. It was particularly rough going — here as everywhere else in the archipelago — because of the strong currents. Yet, I was eager to visit the Baltra canal, because beneath its surface is located a wall composed of one of the most impressive volcanic formations that I have ever had the good fortune to see: a series of hexagonal "organ pipes" of lava, resembling Gothic pillars, in the sea. I had never before seen anything like it; and I have seen nothing similar since that time. In the dim light, it was one of the most haunting and grandiose spectacles that I have ever witnessed. The pipes or pillars were formed by lava dripping from the cliffs of the island into the canal. On land, the cliffs look like a sparkling mosaic of hexagons. Beneath the surface, these hexagons take the form of columns standing on a magnificent carpet of sand.

We also had occasion to witness, in the Baltra canal, a phenomenon with which we had already had some experience in the Red Sea: the cleaning of large fishes by smaller ones who seem to specialize in that activity. There is a very large number of fishes of all species in the Galápagos. That is perhaps why we saw a spectacle here that even the Red Sea, with its own abundance of life forms, was not able to offer. At one of the spots where wrasses relieve large fishes of their parasites, we saw a queue of clients waiting their turns to be cleaned. It was like a crowd waiting to enter a cinema — except that the fishes were quite orderly and waited their turns patiently.

The water in the canal was very clear, and we could often see manta rays swimming past *Calypso*. Some of them were enormous. They swam with their mouths open — mouths similar in shape to the jet exhausts of airplanes. Rays have two small fleshy appendages, one on either side of their mouths, which move slowly as they swim; and their great wings move gently through the water. The manta rays (also called the devil ray) are capable of great speed; yet they move so gracefully that they give the impression of slowness. They feed constantly on the plankton which is so plentiful in the Galápagos. Yves Omer, while diving, saw a school of at least fifteen mantas coming toward him. He says that it was a magnificent spectacle. The rays were on the surface, and Yves was ten or twelve feet beneath the surface. Under each manta there swam a group of fish in pyramid formation. When several mantas swam in a line, all the pyramids were joined at their bases, while the peaks of the pyramids seemed to emanate from the undersides of the individual mantas. The harmony of movement between the rays and the fish underneath was extraordinary. "The mantas were coming toward me," Yves said, "and, at the very last moment, the whole group of pyramids turned together."

Calypso, seen from Gordon Rock (photo taken with our "fish-eye" lens).

Yves has a special affection for manta rays, and he reacts violently whenever he hears tales about the ferocity of these animals. One legend maintains that mantas cover divers with their wings and drag them to the bottom. That, of course, is pure imagination. Most of those who circulate reports about the behavior of manta rays are underwater hunters or fishermen; and perhaps we should not dismiss these complaints too lightly. Certainly, one does not harpoon an animal weighing a ton, and with a wingspread of twelve to fifteen feet, without eliciting a violent reaction from that animal.

Manta rays are quite difficult to photograph, especially when they make the spectacular leaps into the air, for which they are well known, and fall back into the water with a thundering slap. But Yves has succeeded where others have failed. He had heard that mantas always make three successive leaps. As it happened, this was not a legend. He photographed a fine specimen during its third leap — having used the first two to focus his camera.

The Love Life of Sea Lions

I have already had occasion to remark that sea lions are quite common in the Galápagos. There are so many of them that one can see the paths that

Henri Alliet was able to photograph scenes from the love life of sea lions.
Above: The sea lions perform a series of strange acrobatics at the surface. Below: A kiss.

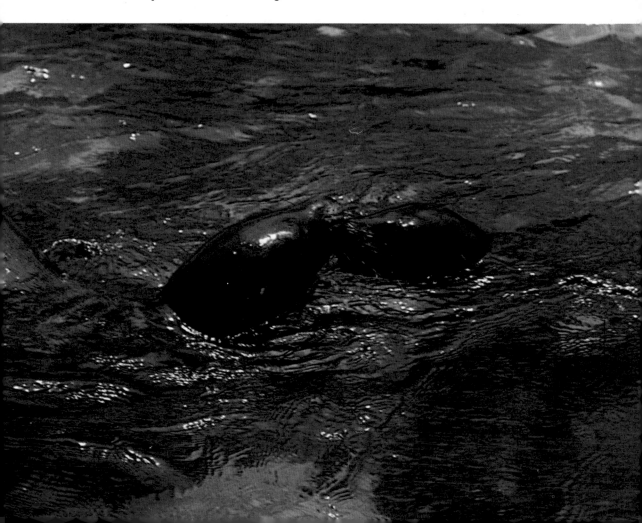

they have worn into the lava as they go from their rocks into the sea. For sea lions always follow the same unvarying path on this short trip.

I have seen many sea lions in my thirty-nine years on the sea. As I have said, we have even had a pair of them as long-term guests aboard *Calypso*. Once, however, in the open sea about thirty miles offshore, I saw a spectacle which I have never been able to forget; an enormous school of sea lions — at least a thousand of them — swimming in tight formation and leaping out of the water exactly like dolphins. I was immediately curious to know the reason such a large group had assembled, and especially to know why they were so excited. I ordered *Calypso*'s engines to be cut and asked our divers to join the sea lions; and our cameramen insisted on participating also.

They soon reported back to us that we had the honor of being the first human beings ever to witness the betrothal ceremony of sea lions. Every year, these animals travel en masse into the open water and choose mates for the new mating season. The young males court the females, and the latter pretend to flee in alarm. There are sexual preliminaries in which the sinuous bodies of the males and females brush gracefully against each other; and there are pursuits and escapes in the depths of the sea. These ceremonies also constitute a tribal celebration of sorts — a feast of love such as was not unknown in pagan antiquity. In this instance, animal gracefulness is enhanced by an almost human tenderness.

An extraordinary encounter in the Galápagos was our meeting with an immense whale shark which must have measured at least forty feet in length. This is an extremely rare animal, but one which we have already seen and photographed in the course of a diving expedition in the Indian Ocean. But we hardly expected to come across such a magnificent specimen in the Galápagos. It suddenly appeared one day between *Calypso* and the coast of Hood Island. I think it is unusual for a whale shark to come so close to a ship, or even to swim in such shallow water. It passed not more than three feet from *Calypso* in water not over fifty feet deep. We launched a Zodiac immediately, but it was too late. The whale shark had already disappeared.

CHAPTER FOUR

Paradise Twice Lost

The Galápagos archipelago is perhaps the place where I have seen the great-est number of animals living together in harmony. They do not fight among themselves. No species seems to prey on another; and it would be impossible to establish the existence of one of those "chain murders" which constitute the alimentary cycle of animals just about everywhere else on earth.

Only the frigate birds give evidence of true ferocity, and they often cause great harm. But here the animals are safe from predators and therefore have never known fear. This is an extraordinary exception in the animal world, for the behavior of almost all animals, as individuals, is dictated by an escape reaction. In the Galápagos, however, animals do not flee, either from man or from other species. Thus, in only one place in the whole world, nature has achieved a state of total disarmament; for the animals do not flee because they are never attacked. There are no weapons, no predators, no conflict; and therefore there is no fear. It is as though nature were trying to prove that the idea of disarmament is not so utopian as some men would have us believe. This is most striking among the birds, which are as approachable as the sea lions and the iguanas. I have already spoken of the irritating boldness of our mockingbirds which were so fearless they never left our men alone.

There are many species of birds native to the Galápagos, and many of them are unusual. It is surprising to find a large colony of penguins here — marine birds which are usually found only in very cold water in the antarctic regions. It is believed that there are approximately fourteen hundred pen-

guins in these islands. Apparently, the cold water of the Humboldt Current is what makes it possible for them to reach the Galápagos and to live here.

Penguins are aquatic birds. Their wings are atrophied and finlike and are covered by very fine, scalelike feathers. These birds feed on marine animals — fish, and especially cephalopods.

The cormorants, a large colony of which lives on Fernandina Island, also have atrophied wings. They are unable to fly; but, when they return to dry land after fishing for food, they spread out their truncated wings to allow them to dry. It is an interesting sight to see these birds, black as the lava on which they rest, standing in the sun with their wings extended like scarecrows' arms.

Cormorants are excellent divers and go far beneath the surface in search of fish and especially of octopus. They have long beaks, which make it possible for them to pull the octopuses out of their holes in the rocks. We would have liked to film this scene; but the cormorants, while they are not at all shy on land, do not like company while they are fishing. We were never able to film one of their underwater hunts.

There are many gannets throughout the archipelago — red-footed gannets which live in trees, unlike the blue-footed gannets and Bassan gannets which remain on the ground. On several occasions, we witnessed a training session of young gannets learning to dive and to handle themselves in the water. Initially, they are very awkward and handle themselves poorly. They seem to be afraid of the water and tend to fly above the surface. Instead of landing in the water feet-first, they either simply drop their bodies into it, or dive with every appearance of extreme reluctance. The older birds, however, never interfere in the learning process. They leave the young birds to train themselves as best they can.

There are pelicans in the Galápagos, and we sometimes found specimens dying of starvation, with their sacs slashed open. In this condition, they are unable to eat, and die slowly and in great agony. They were apparently mutilated by local fishermen, in order to prevent them from fishing — a cruel and even useless act, since the amount of fish taken by pelicans is undoubtedly very small.

In the midst of this abundance of birds, we had to keep reminding ourselves that we had come to the Galápagos not to study these animals, but to make a film on iguanas. It was for the sake of the iguanas that we had gathered special equipment for experiments with these unique animals which

The arid landscape of Plaza Island, with an opuntia in the foreground.

(Above left) A finch. Darwin noted that, in the Galápagos, these birds varied in color from island to island; and it was this observation which inspired his celebrated theory on the origin of species. (Above right) A blue-footed "booby." These birds are very numerous in the archipelago. (Below left) A red-footed "booby." (Below right) A cormorant, a wingless bird, dries its stumps by exposing them to the sun.

A pair of Galápagos gulls.

had never been attempted before. I felt that, with our splendid team of divers, and with several new vehicles that we intended to use for the first time (some new "wet" submarines and new underwater scooters), we would be able to gather information on these underwater saurians that had been unavailable to previous expeditions.

The films made by our two camps already constituted precious documen-

A confrontation between seagulls.

tation; but they were not enough. For almost a century and a half, the iguanas of the Galápagos have intrigued the world of science. What I wanted was to establish once and for all how they swam, how they fed themselves, how long they could remain beneath the surface without breathing, and especially to what depths they were able to dive.

I had asked an eminent iguana specialist, Dr. Bartholomew, to assist at our experiments. On March 20, he arrived at Baltra, where he was met by *Calypso*; and we were then able to begin our first series of controlled observations. I was particularly interested in the results, since one of our chief preoccupations aboard *Calypso* is to study ways of enabling man to become a better diver. We had every hope that a study of the iguana — which is not better suited than man to life in the sea — would give us some useful information on this subject. One of the first questions to be answered, therefore, was to what depth the iguana can dive; and the responsibility for finding the solution to this problem was assigned chiefly to Bernard Delemotte and Jacques Delcoutère.

The diving equipment invented by man can always be improved, but progress in this direction is limited by human physiology. If we are to be able some day to dive deeper, longer, and better, it seems to me that we must look for ways in which to modify human physiology. It was my hope that the iguana — an animal which was not designed by nature to live under water, but which adapted itself to such a life — might yield a clue.

This was a harder task than we had imagined. When a diver attempts to make a marine iguana dive to depths to which it is not accustomed, the animal employs a stratagem: it plays dead. This happened for the first time when Delemotte tried very gently to push an iguana down toward the bottom. At a depth of forty-five feet, the animal settled on a rock and became suddenly still. Delemotte, thinking he had killed the iguana, was very upset. But as soon as Delemotte was distracted, the iguana shot up toward the surface.

In a second experiment, Delemotte succeeded in keeping the iguana at a depth of about sixty feet for a fairly long period. When the animal stopped moving, Delemotte carried it to the surface in his arms. "I started rising, holding the poor animal," Bernard said, "hoping that it would recover. It was spitting water out of its mouth and nostrils. As soon as I put it on land, it revived, and seemed perfectly normal."

Finally, we were able to find an exceptional specimen which quite willingly dived as deep as we wanted it to without ever playing dead.

Unlike human divers, who must have a supply of air, iguanas empty their lungs when they dive. This destroys their buoyancy and enables them to

The "wet submarine" is put into the water.

sink rapidly. Since they have no gills, they are unable to obtain oxygen from the water. Instead, they get it from the tissues of their own bodies. But the oxygen which is removed in this way must be replaced when the iguana returns to the surface. No other land animal that I know of could possibly survive this ordeal.

One of the sailors on Darwin's *Beagle* tried holding an iguana under water to see how long it could survive without breathing. At the end of an hour, the sailor gave up — and the iguana was still alive.

Our own specimen dived to a depth of ninety-three feet. It showed the same symptoms as divers in apnoea — a compression of the base of the thorax, with the stomach becoming extremely flat. It seemed almost an exercise in yoga.

Our New Underwater Vehicles

Aboard *Calypso*, we had some new motorized underwater vehicles which we intended to test and which we hoped would help us to observe the iguanas beneath the surface. These were a new type of scooter and a recently redesigned model of wet submarine. In order to decrease the weight of the diver, we had placed bottles of compressed air on the new scooter and the new submarine. This relocation of the divers' load seemed especially important in the Galápagos, where, in the water separating the islands of the archipelago, there are currents — sometimes conflicting currents — which attain a flow of almost five miles per hour. As one can imagine, it is extremely difficult for a diver to swim against such a current, and I was very eager to see how useful our new equipment would be in solving this problem.

The launching of our wet submarine is a very delicate operation. It is quite heavy, and its plexiglass nose is fragile. The nose, therefore, is installed only when the vehicle is already in the water. The scooters followed the submarine on this initial voyage through the waters of the Galápagos. It was the first time that we used both kinds of vehicle at once.

Only the submarine, which on this occasion was manned by Bernard Delemotte, has controls for turning, diving, and rising. The scooters are guided by the divers, who must turn them in the direction desired. In both vehicles, however, the divers are relieved of the necessity of carrying bottles of air themselves. They need only a mouthpiece, through which they draw air from a bottle attached to the engine.

Delemotte quickly became adept in the handling of the new submarine and especially in the sort of gymnastics necessary to use it effectively. He left it — and his air supply — floating freely, for example, when he dived down to the bottom to inspect a sea fan at the bottom of a cliff. But he had to return quickly to the submarine to prevent it from drifting away in the current and

(Facing page) Penguins, which one thinks of as Antarctic birds, are found in the Galápagos, near the equator.

Land-iguana couples are curiously affectionate.

thus depriving him of his air supply. He devised another solution for an iguana encountered along the way. He brought it back to the submarine and put it in the plexiglass nose of the vehicle, and then took the startled animal for a ride back to *Calypso*, where Dr. Bartholomew and I were waiting for it in order to begin an experiment.

An Electrocardiogram.

Our specimen was an adult male of good size. What we intended was to record and analyze the variations in its heartbeat. The iguana seemed content to let us do whatever we wished. It did not struggle and did not even lash its tail when we strapped it to a table and attached the electrodes, with strips of tape, to various parts of its body. Before returning the iguana to the water, we wished to know what its normal heartbeat was. As it turned out, it was slower than that of human beings: 45 to 50 beats per minute. But, at every beat, three different pulsations were registered by the oscilloscope. The reason was that, unlike mammals, these reptiles have three cardiac cavities and a beat is registered from each of them. The strongest beat, however, emanates from the ventricle; and this is the one that counts in the electrocardiogram.

Now that the normal heartbeat of the iguana had been recorded, we proceeded to continue the experiment in the water. Carefully, the iguana was turned over to Jacques Delcoutère who took it down to the bottom, in the open water off Point Suarez, at Hood Island. Dr. Bartholomew had already taken electrocardiograms, in the laboratory, of iguanas submerged in a tub; but this was the first time — thanks to *Calypso* and its divers — that he was able to take readings in the sea itself.

From *Calypso*, we watched the television monitor as Delcoutère and the iguana went deeper and deeper. Finally, the animal attached itself to a rock. It was still connected by electric relays to the electrocardiograph, which we were watching aboard ship.

By the time the iguana had been in the water for fifteen minutes, its heartbeat had slowed considerably — to 8 or 9 beats per minute instead of the normal rate of 45. Soon, it was down to 4 or 5 beats.

At this point, the iguana halted the blood circulation in its muscles. Its blood was now circulating only between the heart and the brain — the latter being the only organ which requires blood if the animal is to remain alive. The rest of its body was simply cut out of the circulatory system.

Suddenly we saw that the animal's heartbeat had ceased altogether. In such a circumstance, it is easy to conclude that an animal is dead. The iguana, however, is able voluntarily to stop its heartbeat and remain completely without a pulse for up to three minutes without brain damage. Even so, we decided to discontinue the experiment. It would have been cruel to subject the animal to any greater discomfort and risk harming it seriously. Delcoutère rose slowly to the surface, holding the iguana. Gradually, the reptile's heartbeat resumed its normal pace. We were happy to return our diving pioneer to his sun-drenched rock on Hood Island, where he immediately stretched out to begin recovering the oxygen which he had lost and to bring his body temperature up to normal.

Away from Civilization

For a long time, the islands of the Galápagos archipelago were the dream of those who wished to flee civilization and find an unspoiled paradise. Some of these men were not merely dreamers. Some of them actually came to the Galápagos to live. A few of them met tragic deaths; but many of these settlers are still there, living along the coasts of the otherwise deserted islands. They have come from all over the world, and the life they lead is often difficult; but it is nonetheless in accordance with their own wishes and convictions. The population of the islands is about 2,000 — including Ecuadorian officials, the personnel of the Darwin Foundation, and numerous German, Swedish, and French settlers. The largest European colony is at Santa Cruz; but San Cristóbal possesses the most significant historical monument: the letter box of Post Office Bay. It was at Post Office Bay that the privateers, pirates, and whalers of the Pacific came to deposit their mail, and they almost always found a ship to take their letters aboard and carry them to their destination.

Post Office Bay is sheltered, and since it is deep enough to allow easy access to ships, the beach is easily accessible. The famous letter box is nothing more than an old barrel with a small door, and it is no longer used. But it is traditional for each ship calling at Post Office Bay to have its name painted on a piece of driftwood or a plank.

San Cristóbal is also the home of one of the most interesting personalities of the archipelago, Carl Angermeyer. Carl left Germany in 1935, a refugee from the Nazi regime, along with his parents and two brothers, and settled on the island. They were not the first to attempt a settlement on San Cristóbal; but their predecessors had either decided the island was uninhabitable and left, or had died of hunger and thirst. The Angermeyers, however, were able to survive.

I felt an obligation to visit this gentleman. He is very hospitable and

Dr. Bartholomew records the heartbeat of a marine iguana.

The wet submarine, surrounded by divers.

The marine iguana is capable of assuming an air of ferocity. Appearances to the contrary, the spikes of its crest are not sharp.

makes a habit of meeting everyone who visits the island. He may be isolated, but he is by no means a hermit or a recluse. He stays abreast of everything of interest in contemporary affairs, and occasionally he takes a trip to Europe on business.

His business is iguanas. He has managed somehow to tame these animals completely. He began by building his house at a spot with a large iguana population. He made every effort not to frighten them or chase them away, for he was determined to be a good neighbor. As time passed, the iguanas responded by more or less adopting him — so much so that there are almost four dozen of them sharing Carl's house with him.

His house is surprisingly comfortable and well furnished. It is therefore all the more disconcerting to see iguanas everywhere — on the mantelpiece and on the tables and chairs. Angermeyer's innate gentleness and patience certainly counted for a great deal in the taming of the iguanas.

Perhaps the most extraordinary thing about these tame iguanas is that Carl succeeded in getting them to eat food that they do not ordinarily eat: rice patties and tuna patties. But, from time to time, they return to the sea to eat their favorite algae — which is the exclusive diet of the marine iguanas of both Fernandina and Hood Island.

Carl's relations with his iguanas are difficult to define. He is able to recognize several individuals among them by physical peculiarities such as marks on their skin or the size and shape of their crests. But it is not certain that there exists a real affection between the iguanas and the man; there may be instead merely a collective toleration of the human being, and a human toleration of the iguanas. Angermeyer has not named his animals individually. They are all called "Annie" — the name given to them by Carl's great-nephew. But the iguanas share in the life of the family, and seem to like the food that is given them — the bread as well as the fish. This unusual willingness to eat something other than algae is what has made it possible for Angermeyer to take certain of his pets to Europe with him.

Angermeyer's experiences in the Galápagos archipelago — islands so singular in many other respects — were worth studying in depth. I was above all curious to know whether he thought that the iguanas' relationship to him was a purely alimentary one; that is, whether they came to his house only in order to feed. When I put this question to him, Carl's answer was emphatic. "No, no," he said, "though I can't deny that food plays a part in our relationship. That most important thing in a relationship with animals is a man's own feeling toward them. The iguanas have accepted me because, from the very beginning, it was obvious that I wished them well. When I first came here, the iguanas were the only inhabitants. I could have chased them away, I suppose, or even killed them. But I did not do so, and they understood that I was merely a new resident. Together, we have been able to achieve still another example of the harmony among species that seems to be the dominant trait of our islands."

When he showed me the largest of the iguanas, he added: "This one has lived with me for fifteen years. He is the oldest of them. Do you think that he would still be here if I had been mean to him, even unintentionally?"

Angermeyer's house, with its iguanas, its parrot (which belongs to Carl's mother-in-law), and its children, is a center of hospitality for every new arrival to the Galápagos — a place of friendship which may serve as a model for the whole world.

When I left Angermeyer, the fifteen-year-old iguana was lying on the sofa. To me, this seemed the epitome of the spirit of the Galápagos, where man and animal are able to live together in the same environment because

An historic moment: we visit the letter box on San Cristóbal, where the pirates and privateers of yesteryear used to leave and pick up their mail.

each is tolerant of the other and each shows respect for the requirements of the other.

These enchanted isles, which William Beebe, the great naturalist, has called "the islands at the end of the world," are an exceptional place. Men and animals are able to live together peacefully because men are rare, and because the animals, having never been hunted, have no reason to be afraid. But times are changing; and now even the islands of the Galápagos are threatened by an enemy invasion. Here is what Bernard Delemotte told us:

"We had been on Hood Island for about two weeks. Every morning, even though the sea was very rough, we dived and spent several hours being tossed about by the waves and washed up onto the rocks by the breakers. By the time we went ashore, we were usually exhausted, bruised, and afflicted with earaches. We might have been willing to put up with being thrown around, beat up, and deafened, except that, with all that, we were unable to film a single sequence underwater. Moreover, we ran the risk, every minute we spent in the water, of having our cameras and other equipment torn out of our hands and lost in the sea.

"One morning, after trying to shoot for two hours in that incredible swell, we headed back toward land, worn out, with our sinuses pounding. After removing our masks, we glanced around our golden primeval beach — and could hardly believe our eyes. There were about forty tourists — day-trippers, I suppose you would call them — all wearing the standard costume: straw hats, flowered shirts, or bikinis. They were throwing around a rubber ball.

"None of us could say a word. Like robots, we walked toward our camp looking straight ahead. We were all so upset that we didn't trust ourselves to say anything, or even to look at one another.

"We had a silent lunch and, after resting a while, we somehow found the courage to prepare for another dive. While we were getting ready, we watched the day-trippers playing on the beach. A group of them passed in front of us, smiled, and said 'hello.' We were unable to reply. The sea lions had even less patience than we. They had long since fled into the sea."

We discovered later that three tourist boats visit the islands every week. That is not much — but it is more than enough to destroy an animal refuge that we had thought beyond the reach of man's spoiling hand.

Our team in the second camp, on Fernandina Island, tried to handle the tourist invasion in another way and to disguise their disappointment by joking. Jean-Jérôme Carcopino would take two iguanas by their tails, one in each hand; and when he was surrounded by a group of admiring tourists, Jacques Renoir would explain: "He is one of the natives of the Galápagos, and every day he eats two iguanas. Raw. After all, a man needs nourishment."

A Dangerous Situation

It is difficult to be resentful of tourists, since they show an enthusiasm — albeit a naive enthusiasm — for nature. But they are not aware that their very presence may have disastrous effects for this isolated corner of the earth populated by rare and irreplaceable animals.

The relationship between the animals of the Galápagos and human beings is a precarious one. The islands are relatively barren, and food is scarce. Life itself is maintained only with difficulty. All that these rock islands can offer to sustain life forms unique in the world is a tenuous existence. Nonetheless, the islands, up to now, have been something of a natural paradise simply because they were ignored by man. The animals were left alone and had no reason to be afraid. It was one of those rare places on earth which man had only touched; he had never disembarked in great numbers.

It would have been *Calypso*'s greatest adventure if we had been able to discover, on land, an animal world as unspoiled as that of the sea. But this was not to be; for the Galápagos are no longer a remote Eden. Man has arrived.

One of the great contradictions of our time is that more and more people are eager to travel, to visit "the wilds." And yet, at the same time, these people are unwilling to forgo the comforts of civilization. They absolutely refuse to subject themselves to the slightest inconvenience; and, indeed, they are increasingly unable to make the physical effort which life in the wilderness requires. This is the point at which the commercial interests move in to organize transportation and provide all the physical conveniences which the tour-

Carl Angermeyer's house on San Cristóbal. Carl has been on the island since 1935.

ist regards as indispensable and which, although he wishes to travel, he cannot provide for himself. For this reason, we will live to see the day when great hotels will rise in the Galápagos, as they already have at Tahiti, Bora-Bora, and Jamaica, and when the last surviving marine iguanas will live as captives in a concrete pool. On that day, paradise will have been twice lost — lost for man, and lost for animals.

Every day, the free world shrinks a little more. And at the same time, the human race becomes weaker and more dependent. Seeing the tourists on the islands of the Galápagos, I could not help believing that man has degenerated alarmingly. If these men and women had not been led by guides and

A gathering of sea lions and a marine iguana on the rocks of San Cristóbal.

accompanied by a troop of servants capable of providing them with food and finding them a place to sleep, of entertaining them and caring for their well-being, they would never have survived. Our species, I am sorry to say, is physically decadent, and by our decadence we are condemning to extinction other species which, until now, have been free and healthy.

I must say that I am extremely proud of the fact that *Calypso*'s men were able to approach the animals, establish friendly relations with them, and not disturb them in their way of life. I am proud, too, that our men were able to feel at home in their camps on these desert islands. Not only were they able to survive entirely on their own; but they were also able to find peace and contentment alongside the natives of Hood Island and Fernandina — the monstrous dragons of the Galápagos.

An Accident

We spent a total of three months in the Galápagos — three months of navigating *Calypso* through waters filled with dangerous reefs and equally dangerous currents. It would have been a minor miracle if, in all that time, we had never had a mishap.

On the day that I announced our mission in the islands had been completed, we left Hood Island for San Cristóbal in order to pay our respects to the Ecuadorian authorities and thank them for their co-operation during our stay. The sea was absolutely calm, and there was not a breath of wind. The only movement was the Pacific swell.

In the wardroom, we had just finished lunch. Everyone was relaxed, and we had begun comparing notes and impressions on what we had seen in the Galápagos. Jean-Paul Bassaget, our captain, left the wardroom to go up on the bridge. A few moments later, there was the shock of a violent impact. The ship shook from stem to stern. The table and chairs overturned, and several of us were thrown to the floor. I sensed immediately what had happened. We had struck a rock. We were, in fact, in waters which were nothing more than a maze of rocks and reefs. We had passed here before, and Bassaget knew the danger, which was why he had decided to join the watch officer on the bridge.

Aboard *Calypso*, there was absolute silence. I gave a few brief instructions. One team inspected the hold for signs of a leak, while two divers suited up quickly to go down and inspect *Calypso*'s stem.

As it turned out, we had been extraordinarily lucky. Only the observation chamber in *Calypso*'s false nose had struck a rock — a rock which did not appear on our charts. The observation chamber itself was completely

The observation chamber in *Calypso*'s stem was badly damaged when we ran up against a rock which lay just beneath the surface—and which was not shown on our charts.

crushed and had a large V-shaped opening. But it had absorbed the impact and had protected the stem from damage. A few moments later, the team inspecting the hull reported that *Calypso* was bone dry and had not sustained a single leak.

After congratulating ourselves on our good fortune, we set a course for the next stop in the Pacific.

PART TWO
Titicaca

Our two minisubs cross the Andes on the flatcar of a Bolivian train.

Our camera team en route to Lake Titicaca on our special train.

(Following page) We encountered many flocks of llamas. They are strange, proud-looking animals, and we experienced an immediate liking for them.

CHAPTER FIVE

Calypso's Divers Scale the Andes

We are at Crucero Alto, in Peru, at an altitude of 14,920 feet. We are at the heart of the Andes, surrounded by almost vertical cliffs of black and red. Our divers, accustomed as they are to the depths of the sea, now, for the first time, are working at high altitudes. We are at the highest point of our expedition— at almost the altitude of Mont Blanc—and the divers, who are all immune to seasickness, are suffering from mountain sickness. The most ordinary activities, such as bending over or merely walking, cause dizziness and nausea. All in all, it has not been an easy climb.

At first, everyone believed that it would be quite simple; that we would simply sit in a railway coach and watch the view as we rode up to the peak. But before coming to the shunting point, the divers, on several occasions, were obliged to solve crises having to do with the equipment we were transporting. The first thing that happened was that our two minisubs, which were strapped onto an old flatcar, almost went over the side at a turn. It took quick action to save them from ending up in the valley far below. Cables broke, chains gave way, and we had to expend a good deal of energy in rearranging our load—despite the fatigue and discomfort that the least effort entailed.

Our special train consists of only three coaches: an old passenger coach; a flatcar; and a freight car for our equipment—except for the minisubs, which are carried (precariously) on the flatcar. There are seventeen of us—plus twenty tons of equipment, not counting the minisubs.

Our train engineer is so proud of providing us with transportation that he is bound and determined to set a new speed record for the ascent. Whenever the train is going downhill, he opens it up in order to gain enough momentum to climb the next grade in this impossible train route, which follows the ancient road of the Incas across the Andes. The government of Peru has accomplished something of a miracle even in laying a track in these mountains; but it is a track which must, to some extent, follow the contours of the land. It is often so steep, and the curves so sharp, that the entire train cannot maneuver a curve in a single attempt. In some places, we have had to back up and try several times before making it. These, obviously, are the dangerous moments, for despite the fact that each of the cars has a brake, the train is always perilously close to the edge of the cliff.

At Pampa de Arrieros, our train stopped to allow another train to go by on the downward route. This was the first time that we saw the Aymara Indians, whose origins go back to the most ancient Andean civilizations. We stared at them, and they stared back at us. There was great mutual curiosity, but little chance of satisfying it, for our means of expressing ourselves and understanding one another were very limited. We felt separated from them by a chasm as great as that which separates the heights of the Andes from the depths of the sea.

Slowly, our train resumed its progress. On the slopes of the mountains, we got our first glimpses of llamas and alpacas.

These exotic animals, when they are standing still, have a proud and stately look about them; but when they run they seem clumsy. That is deceptive, of course; actually, they are extremely adroit mountaineers. We liked them at first sight, and this impression was to remain with us during our stay in the Andes. We would see many flocks of llamas and alpacas, and we would come to appreciate their spirit of independence. This latter quality, particularly, is quite obvious. Neither a llama nor an alpaca will carry a load of more than about a hundred pounds, no matter what a driver does to him. You could beat one of these animals until it dropped dead at your feet, and it would not take one step forward if its load seemed too heavy. Moreover, there is no record of any man being able to ride a llama. Thus, their usefulness—which originally consisted in providing fine wool to the Andean civilizations—has been rather limited so far as present-day civilization is concerned.

Painfully, the train crawled to the top of a slope. We were up beyond the clouds, and the sun broke through. We were told that the peoples of the Andes used to worship the sun. It is not difficult to understand why. It blazes in the infinite blue sky like an implacable god, bestowing life and light on a landscape of metallic hardness. Beneath us, the clouds billowed in layers, as though the rest of the world was submerged forever in a bottomless sea of vapor.

Having reached Crucero Alto, however, we have at least the comfort of knowing that the worst is behind us. From here on, we have only to go downhill to reach our goal: Lake Titicaca where, for the first time, a diving team will go down into the water—and a diving team with such equipment at their disposal as never has been seen here, at the roof of the world.

An Exciting Departure

The beginning of the line of this incredible railroad on which we are riding is at the port of Matarani, in Peru; and it is there that we began our journey. *Calypso* arrived on October 2, in rather depressing weather. The sea was gray and cloudy, like the sky. The Humboldt Current, which runs along the Pacific coast, cast a chill over the countryside. The swell broke against the rocks like distant thunder, and the rain fell in sheets. Everything aboard *Calypso* was damp, including our spirits. Conversation was subdued and confined to business.

The preparations for our unprecedented diving expedition into the mountains quickly dispelled our torpor, and soon *Calypso* was the scene of frenzied activity. Here are some excerpts from the log of Claude Caillart, who, as *Calypso*'s captain on this expedition, was responsible for those preparations:

Wednesday, October 2.

"We arrive at Matarani at 2:10 P.M. A tugboat circles us, accompanied by a launch carrying a pilot and the port authorities. After signing a dozen papers, I accompany the pilot to the bridge to watch him dock *Calypso*. I ask him not to use the tugboat. He ignores me. The tug smashes into our starboard side.

"Michel Deloire, who had preceded us to Matarani, is waiting for us on the dock. He tells me that our train is scheduled to leave Saturday morning, at 11:30. Gaston, our engineer, is furious. He has been counting on having at least a week to work on the minisubs. They must be lightened considerably for use in the fresh water of Lake Titicaca."

(Above) Lake Titicaca, with its vast expanse and its deep blue color, has the appearance of an inland sea. It is surrounded by bare mountains whose slopes are terraced for cultivation.

(Right) Our camera team prepares for the ordeal of the first high-altitude dive.

(Below) The barge that we used to transport our equipment and the minisubs from Copacabana to our diving sites.

Thursday, October 3.

"Representatives of Peruvian customs seemed to be visiting *Calypso* in relays. They sit down, have lengthy discussions among themselves, drink our wine, and sign great stacks of documents. I have no idea what they are doing, or even what their jobs are; but I make a point of thanking everyone profusely.

"A journalist arrives from Lima and takes innumerable photographs. He tells me that the government has been overthrown and the president ousted. So far, however, there is no official confirmation of this news.

"Everyone aboard is now frantically checking and packing our equipment. The sound of hammers pounding together packing cases drowns out everything else."

Friday, October 4.

"It is 8 A.M., the hour at which we are supposed to begin moving our gear from *Calypso* to the train. It seems that nothing, and no one, is ready. Not even the train. The packing case of a huge piece of agricultural equipment of some kind—a sort of vacuum cleaner for grain, I am told—is protruding over the tracks, and the train cannot reach *Calypso*. It is suggested that I move *Calypso* to another dock; but I protest so loudly that the port authorities decide that the vacuum cleaner is easier to move than *Calypso*. They end up by sawing off a piece of the packing case, and the train is now able to pass.

"Sixteen longshoremen and three chiefs, as stern as justice itself, are waiting on the dock to begin moving our equipment. *Calypso*'s men, after I advise them gently that the time for packing is over and that we must begin loading the train, hand over a few boxes to the longshoremen. Then they give them the cast-iron ballast for the minisubs. The dockworkers do not seem very enthusiastic about this.

"At 10:40, the longshoremen decide that it is time for lunch. I take advantage of the break to slip a few packs of cigarettes to the chiefs. Thereafter, they are more co-operative. I hope that we will be able to leave tomorrow at about one o'clock.

"The afternoon is spent in loading the train and in taking pictures. One of our biggest worries is that the dockworkers, rough as they are, will damage our cameras.

"The minisubs are finally loaded onto the flat car. Our train to the Andes has a most unusual appearance.

"Our preparations are being delayed by a revolution in Lima, either real or imaginary. There is no telephone communication with the capital, and our supplies from that city—packages of food—have never arrived. Late in the afternoon, our Peruvian agent comes aboard and, after a lengthy discussion, agrees to provide us with food and water for the journey.

"With nightfall, a terrible odor of decomposing fish pervades the harbor. Frédéric Dumas says that it smells like something that should go into a good bouillabaisse."

Saturday, October 5.

"Everyone is up early, packing their personal gear. We spend some time sneaking on and off *Calypso,* loading supplies onto the flatcar which were forgotten in yesterday's confusion.

"A little old man wearing a cap—the man in charge of our train—is inspecting the pile of personal baggage in our coach.

"Zoom, our dog, is being left aboard *Calypso.* His grief, when it comes time to part, is touching. We take advantage of it to get some unusual shots of him.

"At 11:30, the train pulls away from the dock. It is a miracle.

"Suddenly, *Calypso* is almost deserted. The few of us left aboard have lunch, and someone notices that Zoom has disappeared. An intensive search discovers him wandering on the dock. His grief has disoriented him. The train on which his friends left was going in the opposite direction."

The train has now begun the long descent which leads into Arequipa Valley. After our journey through a countryside, which seemed dead and entirely deserted, Arequipa resembles a green oasis set in the midst of bare mountains. We can see Lake Titicaca in the distance, an immense blue spot framed by the mountains, lying between Peru and Bolivia—the highest navigable body of water in the world. It covers 3,200 square miles—about the same size as the island of Corsica.

New Techniques

We are oceanographers, underwater cameramen, and divers; but we have abandoned our customary haunts and our exploration of the sea to travel to the highest body of fresh water in the world. Because of the atmospheric and climatic conditions which obtain in this new environment, we have had to work out new techniques. We know very little about the physio-logical implications of diving at high altitudes, and we hope, by means of experiments carried out under careful medical supervision, to answer some of our questions on this subject. But our mission is not limited to this. Not much is known about Lake Titicaca itself—its depth, its geological origins, or the fauna which it contains. These are things with which we hope to be able to deal on the basis of our experience in the sea.

The Andes themselves were the result of continental drift. Volcanic for-mations on the floor of the Pacific acted as an abutment in opposition to the movement of the continent of South America and caused the upheaval which resulted in the rise of both the Andes and the Rocky Mountains. That is why fossils of marine fauna are found in the Andes at altitudes of 12,000 feet, not far from Lake Titicaca. Obviously, the matter from which the Andes are made was once part of the ocean floor.

Another subject for investigation is archaeological in nature. The lake was once the center of intense religious activity. Civilizations grew up around it, and temples were built of which there are still traces. Many of these ruins originated with pre-Incan peoples. Lake Titicaca, for instance, was once the sacred place of the Tiahuanacan civilization. For that matter, it is still a place of pilgrimage for natives of Bolivia and Peru.

A friend of ours—Ramon Avellaneda, who is a diplomat and naval of-ficer as well as a diver—carried out a preliminary exploration of Lake Titicaca. He found much material of interest to archaeologists—especially stacks of stones, which he photographed. We hope to be able to discover submerged ruins of ancient buildings and temples. And, who knows?—we may even discover the golden chain which, as we are told by a persistent local legend, binds together, at the bottom of the lake, the Island of the Sun and the Island of the Moon. It is also said that the Incas, upon the arrival of the Spaniards, threw their treasures into Lake Titicaca.

Less ambitiously, the final part of our mission is to gather information on the depth and topography of the lake. Professor Harold Edgerton, of the Massachusetts Institute of Technology, is supposed to join us in taking sound-ings of the lake with special equipment which he has developed.

All in all, therefore, we have problems in physiology, biology, geology, and archaeology to solve. There will not be much time for sun-bathing.

A Rarefied Atmosphere

Our train stops at Puno station, which is the end of the line. We are on the Peruvian shore of the lake. Our first job is to unload our material and

The colors of the Bolivian flag are taken from those of the flowers around Lake Titicaca.

(Below) Llamas are an integral part of the Andean landscape.

transfer it onto the *Ollenta*, a large British-built passenger boat which was disassembled, carried up the mountains, and then reassembled in Lake Titicaca. It is a strange boat for *Calypso*'s men, accustomed as they are to the somewhat sparse comfort of their own ship. The *Ollenta*'s interior is upholstered in red velvet and finished in mahogany—a luxury boat, no less, hardly intended for the likes of us.

By the night of October 9, everything is aboard the *Ollenta*, and we leave

the Peruvian side of the lake to reach, in the morning, the shore of Tiquina strait, on the Bolivian side of the lake.

Lake Titicaca seems immense—much larger than we had expected. Our reservations about diving at this altitude were well founded. The air is so thin that the cormorants, normally so fast and agile, have trouble taking off from the surface of the lake. Even handling our equipment requires a considerable effort on our part.

We had been forewarned that we were to avoid doing anything strenuous. Only natives of this area are capable of working at this altitude. It takes three generations for a family to adapt physiologically to the rarefied air. The first generation experiences a large increase in the number of red corpuscles in their blood. In the second generation, the thoracic cage develops. And, in the third generation, there is an increase in the secretions of certain glands. These physiological modifications, taken together, allow the human body to carry out normal activities at high altitudes.

In view of these circumstances, we were told that there would be helpers waiting for us at Puno to help us unload our equipment. Unfortunately for us, these helpers never materialized. We had to do everything ourselves. Actually, the work itself did not seem too strenuous, and we were not overly conscious of any special difficulty. The trouble began that evening, when we all experienced the most incredibly painful headaches. Dr. Jacques Tassy took blood samples and announced that each of us was short about three million red corpuscles.

All around Lake Titicaca we can see huge systems of fields and terraces. These are what remain of the irrigated fields and gardens of the Incas, who carried out an intensive cultivation of this area.

Nonetheless, we find the weather uncomfortable. In the course of the first few days, we have been surprised by the cold, which begins as soon as the sun sets in the evening — and even as soon as it is hidden by clouds during the day. These changes in temperature are not only sudden, but quite pronounced. At night it is in the mid-forties, and we all wear caps and sweaters which the natives were happy to sell us. During the day, so long as we are in the sun, the heat is stifling. Our faces and hands have been rather badly burned. But as soon as a cloud passes in front of the sun, or when we are working in a shady spot, we begin shivering with cold.

At Tiquina, fortunately, a barge is waiting for us, manned by two Bolivians and provided through the courtesy of the Bolivian Navy. It is large enough to hold our two minisubs, the compressor (to reload our air bottles), and the seventeen men of the expedition. It moves slowly, since it is powered only by two small motors, but we are thankful for it, and it does get us to Copacabana.

Dr. Tassy was untiring in his concern for the well-being of our divers. One of his chores was to test their blood pressure periodically.

We have chosen this peninsula — almost an island — called Copacabana as our center of operations, since it is located near the site of the sacred islands of the Sun and the Moon.

The countryside here is striking. The lake is surrounded by snow-capped mountains whose bases are gray, black, and red. Almost all the slopes were cleared at one time for crops. I have the impression that this entire area was once heavily planted; but today, this cultivation has been almost entirely abandoned. The dominant color of the landscape is ocher, with black streaks.

At Copacabana we find large eucalyptus trees. And we also see the flower of Titicaca, a red and yellow corolla — the flower from which the Bolivian flag takes its colors.

Our arrival, with all of our equipment and especially our minisubs, has had an impact upon the local population that is difficult to estimate at present. There is a rumor going around that we are going to make use of our machines to desecrate Lake Titicaca which, for the present inhabitants of this area as for their ancestors, has always had a sacred character. The minisubs seem monstrous to them, almost sacrilegious, and they are horrified by our projects and our activities. In order to dissuade us, they tell us all the legends of the lake. They insist that there are whirlpools which swallow up men and sink the light canoes, made of rushes, that are used hereabouts.

One of the things we quickly discovered about the people here is that they drink great quantities of an alchoholic beverage call pisco. A few of our men have tried it and report that it is extraordinarily potent. One of the effects it has upon the people — who drink it out of large white iron jars — is that it renders them capable of intensive and sustained activity. For example, they are able to dance frantically for an entire day and night during their festivals, whereas we are exhausted by the slightest effort.

Delemotte and Deloire are the first to undertake the risk of diving at this altitude. They suit up very slowly, for even such ordinary motions as strapping on their gear and hooking up their air bottles are tiring. In order for us to be able to move and work efficiently here, certain physiological modifications are necessary; but, unfortunately, we would have to wait three generations for them.

We intend to film this first experimental dive, chiefly in order to be able to observe the reactions of Delemotte and Deloire. We suspect that they might exhaust themselves and lose consciousness through lack of oxygen.

Aboard our barge, our air compressor is beginning to have the same problems that we are. The thin air is not supplying enough oxygen to the motor to enable it to start. Dr. Tassy comes up with an ingenious solution: the compressor's carburetor is fed pure oxygen from one of our tanks.

In spite of all our precautions, Delemotte and Deloire have no sooner suited up and strapped on their equipment than they react as though they had suddenly been transported to another planet—one with much stronger gravity than earth. It requires an unusual physical effort merely to bear up under the weight of their equipment. Dr. Tassy records their rate of heartbeat. Their pulse is 95; and it is a strain for them even to breathe through their diving apparatus.

Despite this discomfort, Deloire and Delemotte are determined to continue. They dive, slowly and carefully — and immediately they experience a sense of euphoria. This is easy to understand. At 12,000 feet above sea level, the atmospheric pressure is reduced by half. Every time that a man breathes at that altitude, his lungs receive only half the amount of oxygen that they would get at sea level. But at the same altitude of 12,000 feet, a diver fifteen feet below the surface of the water is surrounded by water at a pressure of one atmosphere, and he breathes a normal amount of oxygen.

Little by little, our divers will adapt themselves to the strangeness of it all, and we will be on our way to solving a very interesting problem in physiology. We want to verify to what degree we will have to correct our rate-of-decompression tables in view of differences in pressure between the atmosphere at the surface of the water and that which exists at a given depth. As it happens, the theoretical calculations that we have already made prove to be correct. But we have also discovered, as the results of our dives here, that a diver in Lake Titicaca at a depth of ninety feet must take the same precautions that he would take in the ocean at a level of 150 feet. In practice, this

(Right) An assistant hands Michel Deloire his underwater camera.

(Below) Albert Falco takes some pottery which a diver found on the bottom of Lake Titicaca.

means that the decompression stops that a diver must make in Lake Titicaca are at ten feet and five feet, rather than at eighteen feet and nine feet as in the ocean.

These tests have been made particularly disagreeable by the alternating intense cold and tropical heat. Also, it is extremely difficult for our divers, wearing their gear, to move about in the open air. It is exhausting for them merely to get into the water and to climb out of it. They have to be helped in and out of their suits; and we have to give them a hand in attaching and removing their air bottles.

From day to day, however, with each new dive by Delemotte and Deloire, who are soon joined by Albert Falco and Jean-Clair Riant, we can see marked improvement in their ability to put up with these strains and discomforts. But we are taking no chances. At the end of each dive, Dr. Tassy examines the divers thoroughly, and interrogates them: "Did it go any better than yesterday? Do you have a headache? Nausea? Are you out of breath?"

The Underwater Landscape

The underwater landscape of Lake Titicaca resembles nothing that we have ever seen in the seas of the world. It is a dark green underwater prairie which extends as far as one can see into the depths of the lake. The explanation for the presence of so much vegetation is that, since the atmosphere is so thin, the rays of the sun are stronger than at sea level and are able to penetrate more deeply into the water, thus favoring the growth of aquatic vegetation.

In the course of one of his dives, Delemotte found that this vegetation grows out of a bottom covered by a layer of mud and slime several yards thick. It is rather liquid, soft, and sticky. This will not make it any easier for us to find anything of archaeological interest. Whatever ruins there may be must have long ago sunk down into the bottom.

Lake Titicaca has the reputation of being filled with fish. Our divers have the opposite impression. They have seen very little fauna.

It has taken us several days to adapt ourselves to the altitude. At the end of that time, however, we find that we are able to work at about the same rate as at sea level — although we must continue to be very careful and to dive only under Dr. Tassy's expert medical supervision. In the water, our divers move about as freely as they do in the sea and are as effective in their work. It is only on land that they find it tiring to move about in their suits.

It seems strange for life on land to require more effort than life in the sea.

Personally, I have suffered a great deal from the altitude. The relief I get while diving seems only to make me more sensitive to the discomfort of moving on land. Sometimes I find it impossible to make the slightest effort. At night, as soon as I fall asleep, I can no longer breathe, and the sense of asphyxiation wakes me up.

The conditions under which we are obliged to work are also difficult. From our base at Copacabana , we have to travel to the Island of the Sun or the Island of the Moon to dive. For this, we have to use our very slow-moving barge, and the trip takes about five hours; so, we have to leave every morning before dawn. We spend the whole day diving, and we return to Copacabana late at night.

We have few comforts at Copacabana; and those we have are due to the efforts of my wife, Simone, and Deloire's wife, Isabelle. These two ladies are in charge of our food supplies, and their responsibility is not an easy one. Food is often hard to come by in this region, and there are seventeen of us to feed. Every day, there is the same problem: the hotel has hardly anything on hand. Simone and Isabelle therefore go out to the local market and buy what they can — which is not a great deal, for the market has little more to offer than the hotel. Nonetheless, they are quite ingenious in getting together a sufficient amount of food to provide us with an excellent lunch on our barge every single day.

The main cause of our discomfort, however, is not the scarcity of food, nor even the thinness of the air, but rather the abrupt changes in temperature. Our noses are now nothing more than a mass of blisters from the sun, and our skin is peeling as though we had never been in the sun before in our lives; and yet, the return trip to Copacabana every night is made in icy blackness and in a wind more cutting than the winter wind in Europe.

After an initial exploration of the lake bottom, we are beginning an archaeological search. Falco, Deloire, Delemotte, and especially Frédéric Dumas and Marcel Ichac, have already started prospecting. It is a difficult undertaking. First of all, the lake is very large, and we are not sure precisely where to begin. We know too that the layer of mud on the bottom, and the algae, will make our work doubly difficult. Everything that the Incas and the pre-Incan peoples may have thrown into the lake is no doubt covered by the mud which has washed down from the slopes of the mountains over the centuries.

Lake Titicaca is contained in a hollow dug by glaciers. Its level has varied with each glacier, and there have even been changes since the beginning of recorded history. Today, it is slowly rising. Therefore, it is not wholly absurd to believe that there are submerged ruins still in existence near the shore or on

Isabelle Deloire inspects one of the pieces found in the lake.

the sacred islands. And so, we have begun exploring the areas near the shores for structures which may have survived beneath the surface of the water.

Our digs beneath the surface are very hard work for the divers. They must dig in mud several yards deep; and of course the water immediately becomes so cloudy that they can hardly see what they are doing. But there are rewards. Earlier today, Frédéric Dumas and Jean-Clair Riant were digging away when they suddenly turned up some pieces of pottery in the mud. These first finds consisted of kettles blackened by smoke. They were of comparatively recent origin, but they were intact. It was an interesting discovery, but we found no painted vases or jewels in these first efforts at excavation. Yet, the legends of this place assure us that such things are buried here in abundance.

We have now completed our study of the physiological aspects of high-altitude diving, and we are satisfied with the results. I have a feeling that equally satisfactory results will be harder to obtain in our archaeological project. We have found some stones, placed in a pattern, on the bottom toward Tiahuanaco. They are similar to those described by our friend Avellaneda, but our divers have been unable to dig up any artifacts around these enormous stones, which were either walls or were so placed as to outline a rectangle. It must be that the mud has swallowed everything.

CHAPTER SIX

The People of the Rushes

Copacabana, our headquarters during the six weeks that we will remain at Lake Titicaca, is the most famous place of pilgrimage of the Bolivian Indians. A sumptuous church, covered with massive statuary, is visited by hundreds upon hundreds of faithful from the entire Andes region. Many of them make the journey from La Paz on foot, over a hundred miles of poorly defined paths.

The town of Copacabana reminds me both of Lourdes and of Deauville. But it is a Deauville without a casino and often without water. Electricity is available only three hours each day, and there is never any heat in our hotel.

My wife, Simone, has an excellent eye for local spirit and customs. The following passages are part of her observations on the religious beliefs of the natives of this area:

"The Indians come to pray to the black Virgin in a cathedral built in the most atrocious style, but absolutely crammed with treasures. . . . The Christian fervor of the pilgrims is mingled with pagan beliefs. They are a people with a long history of oppression from many sources. Not surprisingly, their idea of a pilgrimage is to obtain protection against bad luck and against all the possible troubles which are the lot of a poor Indian in the heart of the Andes. To them, a blessing by the priest is a universal panacea; thus, they have him bless everything they own — their trucks and cars, their motorcycles and bicycles. The trouble is that these vehicles are usually so dilapidated that the services of a good mechanic, rather than those of a priest, are called for.

"On Sundays, after High Mass, the priest, accompanied by two altar boys, goes out into the square in front of the cathedral. With a great deal of ceremony, he bestows his blessing on every heap of scrap metal that the people have been able to drive, pedal, or push into the square. When the benediction is over, firecrackers explode, confetti falls from the rooftops, and streamers fill the air. The owners of each vehicle — the godfather and godmother, one might call them — then baptize their old American or Japanese vehicle with beer. There is much shouting and much embracing in the Spanish style. The final phase of the ceremony consists in drinking the health of the black Virgin, and of the vehicle, with pisco. Not infrequently, there is a sequel to the ceremony: the newly blessed and baptized car or truck ends up, at some point during the night, smashed against a tree or at the bottom of a ravine."

Our arrival in Copacabana, as I have already mentioned, was accorded a rather strange reception in this town, where pisco and the black Virgin are regarded with equal reverence. To the Indians, we seemed strange beings — perhaps even supernatural beings — whose purpose in coming was to destroy, or at least to offend, the spirits who dwell in Lake Titicaca.

Our minisubs were regarded with even more awe than we. It is not unlikely that the minisubs — which we also call diving saucers — seemed to them either something out of the ancient magical lore of their people, or else something out of the science fiction of the present age. It seems probable that there was some confusion in their minds between diving saucer and flying saucer; and this led to the suspicion that we were men from another planet.

In order to allay these suspicions, and above all to give a present to these people of one of the festivals which seem to be the only bright points in their lives, we decided to have our minisubs exorcised and blessed. Having arrived at this decision, the next step was to appoint godparents to the two machines. Albert Falco and Gaston Roux seemed the most likely candidates, for they are, respectively, the pilot and the engineer of the minisubs.

The festival itself was organized, with great enthusiasm, by Simone and Isabelle. They emptied the shops of firecrackers, confetti, and streamers. They even managed to obtain the services of the fireworks expert of the area, who was highly esteemed for his great experience and much in demand in all the villages for every festival. I must confess that we experienced a twinge of doubt concerning this gentleman's competence when we first saw him. He had only one eye. We did not dare ask how he had lost the other.

"After much shopping around," Simone recalls, "we found some ribbons in the colors of France (blue, white, and red) and of Bolivia (red and gold). We had decided to decorate the diving saucers exclusively in the colors of the two countries.

"By Saturday, October 19, everything was ready; and everyone knew that those mysterious Frenchmen were going to give a great festival to which everyone was invited. And, it was said, there would be much free pisco to drink. For these people, trapped in a bleak and barren land, deprived of all entertainment other than what they can create for themselves, the unexpected festival was a gift from heaven.

"We were able to enlist the services of nine different groups of dancers, singers, and musicians. There was even a flock of nineteen llamas invited to participate. The priest of the place, delighted at the trouble we were going through for his people and enchanted with the idea of being asked to bless the minisubs, promised us a surprise of his own for the festival. And, he added, if it became necessary to repeat the ceremony of benediction several times in order for our cameraman to get good shots of it, he would be more than happy to oblige.

A Variegated Crowd

"The next day, Sunday, at the crack of dawn, Isabelle and I went out to the dock to decorate the minisubs. It was freezing. And it may have been the cold that addled us, so that we made a serious diplomatic blunder. We decorated the first minisub in the colors of France. Then it was loaded onto the back of a rickety truck and driven through the streets of the town to the main square. The sight of it created a great stir, and not a very favorable one. It was not our taste in decorations which offended the people, but our choice of colors. The blue, white, and red of France seemed an insult, and a deliberate one, to the sensibilities of these proud Bolivians. As the minisub passed through the streets, women ostentatiously tossed flowers at it — the flowers which they call *cantatos* and which are red and gold, the colors of Bolivia. Finally, when the minisub was unloaded in the square, a lady of dignity, wearing a melon-shaped hat, thick black braids, and a wide skirt, approached the machine and solemnly attached a Bolivian flag to its antenna. At that instant, happily, the second minisub arrived, decked out in ribbons of red and gold. The crowd made noises of approval. I detached the flag from the first minisub and handed it back to the lady. She seemed embarrassed.

"There were already many people in the square, milling about, talking,

(Following page) One phase of the celebration held upon the occasion of the blessing of the minisubs. In the foreground is a local musician playing Pan's flute.

dressed in many colors. The women gossiped together, their babies strapped to their backs. It was a gorgeous day for a festival.

"Suddenly, out of the streets leading into the square, the hired groups of entertainers began to stream into the crowd. The musicians came first, wearing enormous feathers, or sombreros glittering with paste jewels. The noise was deafening. The cymbals clanged, the drums boomed, the trumpets blared and the flutes screeched. Each musician seemed to be playing for himself alone, without bothering about what the other members of his group were doing. To add to the confusion, the flock of llamas now entered, heads high, haughty, disdainful, and expressing their contempt by allowing their bowels to move freely.

"At precisely ten o'clock, the priest appeared at the main door of the cathedral between two altar boys. He was wearing a cope of silver lamé. This was the surprise he had promised. As the priest moved, the cope raised a cloud of golden dust which settled on the sellers of holy candles, who had piously dropped to their knees.

"Silence fell upon the square. The instruments were quiet. Only the songs of the birds on the roofs could be heard.

"The priest, his eyes downcast and his hands joined, spoke the Latin prayers of the rite over the minisubs and sprinkled them with holy water. Then, raising his arms toward heaven, in a solemn voice he called down God's mercy upon these complex machines standing incongruously before his cathedral, surrounded by his flock, 12,000 feet up in the snow-capped Andes.

"The religious aspect of the ceremony was over, and the priest returned to the cathedral. The bells began to peal. A storm of confetti descended upon the square, while fireworks exploded and streamers streamed. The crowd was so dense, and so enthusiastic, that Isabelle and I barely managed to baptize the minisubs with the traditional container of beer. Then the music began again, and the dancers whirled through the crowd. The matrons of the town came forward to kiss Albert Falco and Gaston Roux, the proud godparents; and then they came to Isabelle and me and kissed our hands. Michel Deloire was filming away. And Marcel Ichac was shouting orders which no one heard."

The costumes worn by the Indians on this occasion are reserved for use during the festivals which are the only joy of their bleak existence. These costumes — like the music of the people and their songs — have no doubt been influenced both by the Incas and by the Spanish conquerors; but it is difficult to tell where the influence of one ends and that of the other begins. Certainly, the flutes, drums, trumpets, and horns are of purely Indian origin; but all of the stringed instruments date from after the Conquest. They were unknown

(Left) Marcel Ichac and Simone Cousteau, who helped organize the festival in honor of the minisubs. (Right) The priest of the church at Copacabana arrives with his altar boys, to bless the minisubs.

in the Andes before the arrival of the Spaniards.

The wide, pleated skirts of the women are obviously of Spanish origin also, even though they usually contain the colors of Lake Titicaca, red and gold. But we have no idea of where their hats originated — those great melon-shaped hats of which they are so proud and for which they spend a relatively large amount of money. Actually, they are more than hats. They also serve as drinking containers, and as buckets in which to carry water, to wash dishes, and to scrub floors.

Just about everywhere that we go in the countryside around the town, we see women wearing these hats, stooping, their voluminous skirts wrapped around their legs. It is not difficult to guess what they are doing. They are not at all embarrassed, but greet us with a smile. After all, they wear seven or eight skirts; but no pants.

The faces of the Aymara Indians are very wrinkled, sunburned, and often have large brown patches. The people have an oriental, rather than an occidental, appearance, with high cheekbones and flat faces. There is no longer much doubt among anthropologists that the people of the Andes are of Asiatic origin.

Both men and women drink to excess, their favorite beverages being grape juice mixed with various alcohols of more or less unsavory aspect; and, of course, pisco. They are also much addicted to the use of drugs, in the form of the cola nut, which they chew constantly. Debilitated by the altitude and

(Above) A local musician at the Copacabana festival.

(Right) One of the masked dancers at the celebration.

by alcohol, they are nonetheless capable of hard work. They carry heavy loads on their backs, supporting them with a strap around their foreheads.

For food, they raise crops in small fields on the slopes of the mountains. These slopes have been terraced for this purpose since the time of the Incas. Their crops are the traditional ones, which have been handed down for thousands of years from pre-Incan civilizations: corn, and a species of potato which is the ancestor of that known to us. The land is fertile and competently irrigated.

The llama and the alpaca furnish the Indians with wool, which they weave into bedcovers, ponchos, and, above all, into the striped caps which we have adopted as our official headgear at Lake Titicaca.

A Forced Dive

The blessing of the diving saucers was the occasion of a day of leisure and also of a somewhat difficult test for our machines. At about noon, after

the religious ceremony had been completed and the secular celebration was well under way, I returned to the hotel for a rest. I had been there only a short time when someone came to tell me that a procession, led by the musicians, had moved the minisubs down to the dock, and was now demanding that they be launched immediately. Most of the town's population had assembled to watch the proceedings, and they were determined not to be disappointed.

Unfortunately, the launching of a minisub is not as simple as the town-folk, under the inspiration of pisco, seemed to think. Nonetheless, we felt that we owed them a demonstration. I therefore dispatched a truck with a hoist out to the jetty. Raymond Coll climbed into the minisub, and it was then lifted by hoist and lowered into the water. It sounds simple, but I was very dubious about the whole operation. And Raymond Coll was even more dubious than I, although I strongly advised him to pray to the black Virgin.

The fact was that the minisub was not ready for fresh-water diving, and if the cable had broken it would have dropped like a rock to the bottom of Lake Titicaca. He had planned to install special floaters, but they were not yet in place.

Luck was with us. The cable held, and the minisub was "launched" to everyone's satisfaction. The festival could now proceed unclouded by doubt.

We are quite surprised at the intensity of passion and even of faith which the Indians showed in their dancing. Each one represented a personality, and each one played his assigned role with great conviction. Some of them rode on horses of cardboard decorated with forks and spoons, or any other shiny object. Others transformed themselves into a monster of straw and attacked the shrieking crowd. Most of the people were wearing masks.

The excitement of the crowd was beyond belief. When it seemed to us to have reached its apex, there was some new provocation and even greater excitement. The bands played in relays; but the dancers, apparently in a trance of some kind, never ceased their whirling. The strength of these people under the effects of alcohol and the cola nut is astonishing. Monday morning, at six o'clock, the dancers were still dancing. Shortly afterward, however, the festival was officially over, and little groups of people began leaving for their own distant villages, led by flutes and drums.

The Fauna of Lake Titicaca

The divers have now begun an in-depth exploration of the lake, going down as far as they are able. They have come to look forward to these visits to the bottom, which make it possible for them to breathe at a normal pressure.

The people living around the lake are astonished by our audacity and fearful for our safety. They believe that the sacred lake is inhabited by devils and other mysterious beings, and that it is a place of supernatural manifestations.

A number of natives have told us that if we dive at certain locations, we will hear the sound of bells ringing, or find submerged ruins. On each occasion, we dived at the designated spot and found neither.

These dives have not always been agreeable experiences. The water is cloudy because of the mud, which rises as soon as it is disturbed. Sometimes the divers go through veritable forests of algae so dense that they must push it aside in order to be able to pass. The presence of this vegetation and of the mud on the bottom has added to the difficulty of our mission.

In sixty feet of water, Falco found the remains of a small steamboat which sank, we are told, in 1942. Such accidents are not rare here, especially because of the sudden storms which rise on the lake. The hull of the boat was probably six or seven feet high, but only eighteen inches now show above the bottom. The rest has been swallowed by the mud.

The divers quickly noticed that this wreck was inhabited — by giant frogs. These were the first such specimens that we have seen, though we had an inkling that we would run into them sooner or later. These frogs are unique, and, in 1927, they were the subject of an intensive study by an American expedition.

This species of frog (*Telmatobius culeus*) differs from other species drastically. All other frogs are obliged to return to the surface regularly in order to breathe. Those of Lake Titicaca would suffocate if they were exposed to the air. Their lungs have atrophied to the point where they are useless. Yet, they do not have gills, like fish. Instead, they breathe through their skin. The tiny capillaries which cover the surface of their bodies draw oxygen directly from the water.

These frogs spend most of their lives on the muddy bottom of the lake. When they are disturbed by divers and obliged to move, one can appreciate their full size. They are at least twenty inches long.

It is interesting to speculate as to why this particular species of frog is able to remain beneath the surface and breathe in the water. It is likely, I think, that the thin air of the Andes simply did not supply enough oxygen to their ancestors, and, as they grew in size, their lungs shrank. Eventually, they adapted to their environment in the only possible way: by becoming capable of breathing through their skins and of removing oxygen directly from the water.

Living in the water as they do, the frogs have become mute. Even during

abundance in the lake. It is called *totora* by the Urus, and they use it to build handsome boats of which both the stems and sterns are woven to a point. Even their sails are made of rushes.

Totora is also used in the construction of the Urus' huts, and it is the material from which their "floating islands" are made. In the latter case, layer upon layer of the rushes are stacked one upon the other until the island attains a sufficient mass to serve as a base for huts. When the top layers begin to decay, freshly cut rushes are piled on top of them.

The tubers of the totora, or *kzouras*, are eaten by the Urus. They also provide them with a fermented beverage. The tribe eats cormorants, which they have domesticated and raise as other people raise chickens.

The Urus are a superstitious people; the last vestige of an ancient civilization. This gave us an idea for an experiment — or rather for a practical joke. Jean-Clair Riant, wearing his diving gear, sneaked under a floating island; then, breaking through the rushes, he jumped up in the middle of the Uru village square. With his mask, air bottles, and other gear, he could well have passed for a monster risen from the sacred lake. But the Urus were more sophisticated than we had thought. They simply glanced at Jean-Clair and went back to whatever they had been doing. Jean-Clair did the only thing he could do — he left as quickly as possible. It may be that the Urus were having the last laugh, for when Jean-Clair emerged from the water we had to give him a thorough scrubbing. The underside of the floating island, as it turns out, is literally a cesspool and garbage dump, and our friend returned exuding the most incredible odor to which it has ever been my misfortune to be exposed.

There are many interesting aspects of Uru culture. The people are born, live, and die on the water, and yet they are deathly afraid of it. They are convinced that if a man falls into the lake, he will die instantly. Also, they share the belief of many primitive peoples that a camera is an instrument of black magic and that it takes possession of a man's soul as well as of his face. Jean-Clair Riant's surprise visit to their village, however, provided a moment's distraction for them — just enough time for us to film their village with relative freedom.

The Urus are the most ancient inhabitants of this region of the Andes. They antedated not only the Incas, but also the Aymaras, on the shores of Lake Titicaca. They are considered to be the last survivors of the original inhabitants of the American continent.

(Following page) Jean-Clair Riant rises to the surface in the middle of the Uru village.

According to some authorities, the Urus have many qualities in common with certain Amazonian tribes. Their language, Puquina, is related to Arawak — one of the principal languages of tropical America. It is believed that, at one point, the Urus left the basin of the Amazon and moved into the mountains; but no one can guess at the reasons for this supposed emigration.

Today, the Urus number no more than two hundred, and even those few survivors are not pure-blooded Urus. There are a few at Irou Itou, a village on the shore of the Desaguadero river, twenty-five miles from the place where the river flows from Lake Titicaca and begins its southward course toward Lake Poopó.

The Urus were undoubtedly on the decline at the time of the Incas. They were not an agricultural people, but lived by fishing. It is believed that the Incas refused to initiate them into the cult of the Sun, but confined themselves to demanding tribute of fish and baskets — the latter being a commodity which the Urus produced with unusual skill.

Our relations with the Urus did not begin and end with Jean-Clair Riant's unexpected visit to their village. We made an effort to establish communication with them. This, however, was difficult; for even though there is a school in their village, very few of the inhabitants speak or understand Spanish. Only the children could understand us sufficiently well to serve as interpreters — a tribute to the progress made in this school.

The Uru language is spoken only by a few hundred people today, and this lack of ability among the Urus to communicate with other people has had a fatal effect upon their development. They seem to have nothing left of their ancient civilization.

We tried to question the chief of the village, but he seemed unable to rouse himself from the dreamlike trance created by pisco and the cola nut. Or it may have been that his silence was the last refuge of a wise man. I am inclined to think so, for the floating islands of the Urus are now menaced on every side by tourists. This last vestige of an ancient people has become an object of curiosity to strangers who come to photograph them and to buy the few objects they make — little terra cotta figurines, hides, and especially woven objects such as baskets and models of their rush canoes. In this, at least, the Urus have retained a trace of the skill of their ancestors.

The coming of the tourists undoubtedly constitutes a danger to the Urus. If they are deprived of their isolation, they will probably lose the last remnants of their unique culture and their skill in handiwork — a skill which was once famous. It may be that this loss is compensated for by the prospect of healthier, and even more comfortable lives for these Indians.

From the depths of Lake Titicaca, a diver brings back specimens of the lake's frogs. The lungs of these animals have atrophied, and they breathe through their skins.

The frogs are placed in plastic bags for later examination.

The diver makes a last check of the minisub before it dives beneath the surface of Lake Titicaca.

(Left) Section of the lake bottom explored by the minisub. (Right) Soundings taken across Tiquina Strait.

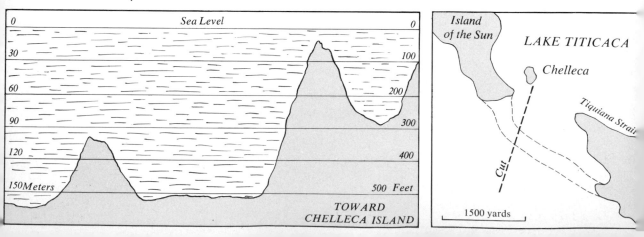

CHAPTER SEVEN

The Golden Chain

Most of the inhabitants of Copacabana, and perhaps all of them, were convinced that our purpose in coming to Lake Titicaca with our elaborate equipment was to find "the golden chain."

A persistent legend maintains that the Incas, or their predecessors, joined together the Island of the Sun and the Island of the Moon with a chain of gold beneath the surface of the lake. There was also said to be an immense disk of gold somewhere on the bottom of the lake. In the minds of the people, only the hope of discovering these treasures could have motivated us to work as hard as we did at 12,000 feet.

It is very likely that, if we refused categorically to look for the chain, the people would have been extremely disappointed, for they had come to admire our courage and our efficiency in the water. So, since hunting for the chain involved nothing more than exploring the bottom of the lake, which we had every intention of doing in any case, I saw nothing wrong with including the area between the Island of the Sun and the Island of the Moon on our agenda. Who knew what we would find? There might not be a chain of gold — but there might be other things of less commercial value and of greater significance.

What the people did not realize was that, even if the chain had existed, we would never have found it. It would have been swallowed up long ago in the muds of the bottom of the lake, and it would have required incredible luck on a diver's part to be able to locate it. As it was, every time one of us stirred up the mud, the water became so cloudy that we had to wait almost an entire day for it to clear again.

Another obstacle to any search of this kind was the thick vegetation growing on the bottom in certain parts of the lake. These dark aquatic forests were not composed of totoras, but of fresh-water algae, which were both soft and dense. It was difficult even to get through a patch of this growth and almost impossible to reach the bottom wherever this algae grew. A diver in the midst of the algae was cut off from all light. It was as though he were in a closed box. Sometimes his mask and his mouthpiece became entangled and were ripped off. And it was easy to lose one's sense of direction and not be sure of which way was up and which way was down.

It seemed to us that there was no fauna at all in these stretches of algae. In any event, we saw none. But it is true that it was almost impossible to see anything — let alone a chain of gold.

Dr. Edgerton's Arrival

Between the islands of the Sun and the Moon, there is an underwater valley into which our divers could not go because of its depth, which was over 300 feet.

Our explorations of the bottom of the lake were facilitated by the arrival at Copacabana of Professor Harold Edgerton, who came to join us as he had promised. With him, he brought a modified sounding device whose acoustical signal indicated not only depth, but also the nature of the bottom (mud, sand, or rock).

This sounding device is relatively small and easy to handle. It was extremely useful to us in establishing the contours of the bottom and in choosing the area most suitable for a dive with the minisubs.

The range of the Edgerton sounder gave us a detailed profile of Lake Titicaca. The maximum depth, as it turned out, was 1,235 feet. This was a far cry from the fantastic depths of over 3,000 feet which natives had assured us were usual in the lake.

Dr. Edgerton explained his findings to us in detail, and, after a long discussion, we decided on an area which seemed particularly interesting for exploration in the minisub. According to the sounder, not far from the islands of the Sun and the Moon there is a basin of sediment about 400 feet deep located between two rock formations. From that flat bottom rise steep cliffs, which will probably be relatively easy to inspect.

Moreover, the famous golden chain of the Incas is supposed to be located in this same area. Thus, we will kill two birds with one stone.

The reason we had taken so much trouble and expended so much effort in bringing our minisubs to Lake Titicaca was to use them to explore the lake, the deeper parts of which are totally unknown. It was not enough for us to ask Edgerton to take soundings. I wanted to have a visual record of the marine landscape which has given rise to one of the great legends of the New World.

Our minisubs, however, were designed for use in the sea. They could not be used, without modification, in Lake Titicaca. The reason is that an object is less buoyant in fresh water than in the ocean. We calculated that the minisubs were about 150 pounds too heavy to be used in Lake Titicaca. We would have to counter that weight — or they would sink to the bottom. I had already had a system of rigid floaters made aboard *Calypso*, and we had fastened them to the upper part of the minisubs.

Another problem was the launching of the minisubs. Aboard *Calypso*, we have a special hoist which does the job in a few minutes. Here, however, it seemed we would be able to launch the vehicles only by loading our truck hoist onto our barge and taking it out on Lake Titicaca, into water deep enough for the minisubs. We would not have been able to use the minisubs, therefore, if it had not been for the help of the people of the village of Yampopata, the community nearest to the Island of the Sun. In forty-eight hours, they constructed a rock jetty into the lake. The water at the end of the jetty was deep enough to launch the minisubs — and the truck hoist could reach that spot simply by driving out onto the jetty. So the vehicles were launched without too much difficulty.

It turned out that we had been correct in our calculations, and the minisubs floated very well in the lake, because of the perfectly balanced floaters. Albert Falco took the controls of one of the vehicles, and Raymond Coll climbed into the other.

"The hatches were closed, the last checks run, and we were in the water", Raymond Coll told us. " It was hard to believe that we were beneath the surface some 12,000 feet above sea level rather than in the Mediterranean on a sunny winter day. This was also Falco's feeling, as he explained to me on the ultrasonic telephone.

"Divers accompanied our minisubs to a depth of 100 feet to make sure that everything was all right and that they had not been damaged during the train trip through the Andes."

(Following page) The minisubs and our team aboard the barge, *en route* to Tiquina Strait.

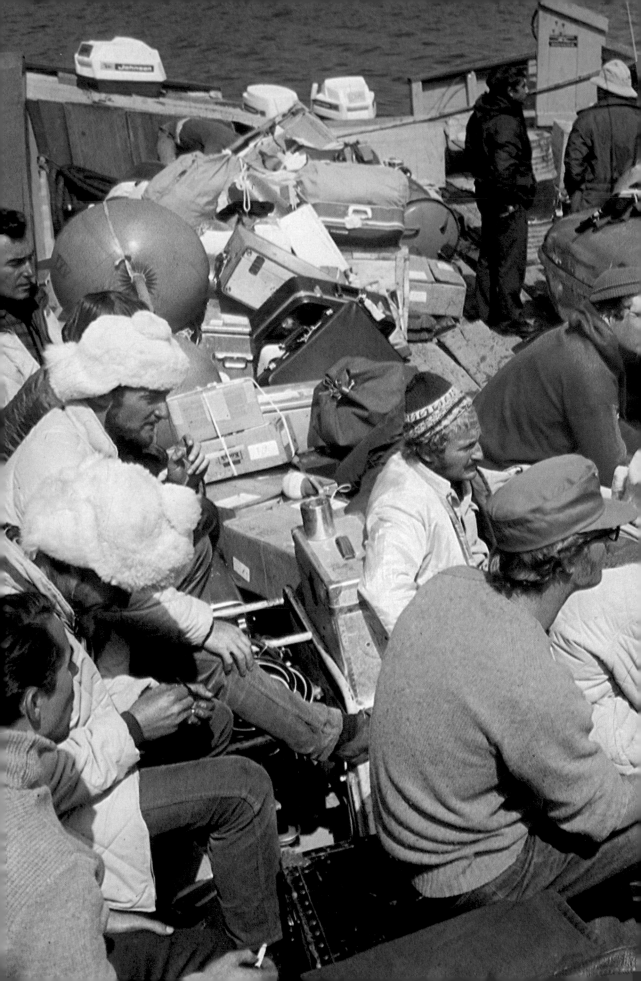

In spite of the necessary modifications made in the minisubs, they were just as easy to handle as they had ever been. The pilots felt at home, surrounded by familiar instruments, comfortable — perhaps for the first time since reaching Lake Titicaca.

At 100 feet, Coll blinked his lights at the divers — our traditional method of saying *au revoir*. Then Falco and Coll went down deeper into the lake.

Each of the minisubs carries two cameras, one for close-ups and one for far shots. And they are both able to operate at depths of up to 2,000 feet — a depth which does not exist in Lake Titicaca.

The guides on this dive were the new charts drawn up for us by Dr. Edgerton. At 375 feet, the minisubs reached the bottom of the lake and followed the rock spine which Edgerton had found with his instruments and delineated on the charts. The water was almost as clear as that of the sea. Fresh-water plankton, it seemed, was somewhat less abundant than its salt-water counterpart.

Falco and Coll were both sensitive to the fact that what they were seeing had never before been seen by man. The bottom of this lake, up until their descent, had remained more inaccessible than the surface of the moon.

Once they had crossed over the rock formation, they began their long journey over the mud at the bottom. They had wanted to explore the sedimentary basin located between Copacabana, the Island of the Sun, and Cheleka Island. According to Edgerton's computations, the layer of sediment accumulated at that spot would be almost thirty-five feet thick.

When they reached the middle of this plain, they noticed that there were strange markings in the mud of the bottom. Coll told Falco about it, and they decided to follow the markings, which seemed to follow a regular zigzag pattern. The mystery was quickly solved. They caught up with two frogs moving along the bottom. The markings were caused by their wake. They were quite surprised at the sight of these animals, for frogs usually cannot live in water this deep.

This plain is bounded by a rocky ridge, also duly noted on Edgerton's charts. It resembles a giant spinal column lying along the bottom.

By the end of this exploration, the two men had gathered valuable information on the composition of the lake's bottom. But, even in the bright glare of the minisubs' headlights, they saw no trace of the golden chain of the Incas.

As soon as the minisubs reached the surface, they spouted like whales. They were designed to do this in order to make them easy to find in rough water. On this occasion, their friends were waiting for them and found them easily.

A minisub surfaces near a native canoe.

The local population turns out to greet our divers.

The Indians also found them, with equal facility. Unable to restrain their curiosity, they had rowed out to watch the surfacing of the two underwater vehicles which appeared to them to be both diabolical and wonderful. The sight of the minisubs — the latest word in marine technology — side by side with the primitive canoes of the Urus, was a perfect summary of the history of Lake Titicaca over the past two thousand years.

The people of this area have a great love for drawing, painting, and embroidery. It is likely that our minisubs have now been incorporated into decorative themes, and that they are seen not only on walls but also on the Sunday shawls of the women. If so, we have unwittingly provided new subjects for the folklore of the Andes.

The Mysteries of Tiahuanaco

The recent explorations made by Argentine divers, and especially by our friend Avellaneda, had led us to hope that we might find something of archaeological value in Lake Titicaca. We were particularly interested in this aspect of our mission since the lake played a significant role during the Tiahuanacan civilization and the Incan period. During the final centuries of the pre-Columbian era, there was a cult consecrated to the god Viracocha, who had emerged from the waters of Lake Titicaca. And Koati Island was believed to be the home of the moon.

Snakes played an important part in this cult, for the snake was thought to be the enemy of lightning and thunder. This belief has persisted among the present-day Aymaras. We saw sculptures of serpents on the walls of their houses.

It was also on one of the islands of the lake, in about A.D. 1100, that Manco Capac, first ruler of the Incas, had a revelation concerning the part planned for him by the sun god. The Incas built a magnificent temple on the island, the walls of which were covered with gold. Every Inca was expected to visit the temple at least once during his lifetime and to make an offering of precious metal.

We paid a long visit to this Island of the Sun; and we even made use of a metal detector, though without much hope of turning up anything. But we were not terribly interested in gold. What we had really hoped to find was pieces of pottery; and we looked for relics of this kind during our dives around the island. All that we were able to discover, however, were some very modest ceramic pieces which were probably household items from the period of the Incas.

I have already described the general aspect of the lake bottom, where patches of algae alternate with stretches of deep mud. This composition makes it very difficult to find anything on the bottom. It is likely that the mud has hidden forever anything that lay on the bottom — and especially the gold that the Incas are supposed to have thrown into the water when the Spaniards arrived.

It would be fascinating to undertake an archaeological dig here, but it would have to be done with equipment much more powerful and sophisticated than ours. For one thing, the mud would have to be dredged out and then examined inch by inch for its contents. It is almost an impossible task — if for no reason other than the size of Lake Titicaca.

We made a special effort to dive near the submerged stones found by our friend Avellaneda. These are piles of blocks and structures of one kind or another; but it seems to me farfetched to see in them the ruins of cities or even temples. I am more inclined to believe that these stones were used as a bulkheading or sea wall, to protect ships against storms on the lake.

The stones lay at a depth of between fifteen and twenty feet. Twenty years ago, they were at the level of the surface. Obviously, the water level of the lake has risen or fallen in different ages, for the lake is fed by water from the mountains — and this is what has carried sediment and mud into the lake.

In the neighborhood of Oré, we found an imposing ruin made of well-joined stones. This structure seems to be the subbasement of a building of

We were joined at Lake Titicaca by an old friend, Professor Harold Edgerton, of MIT. Professor Edgerton took depth soundings of the lake with some new instruments which he had developed.

Jean-Michel Cousteau, who was in charge of the logistical aspects of our expedition, went slightly native—at least to the extent of adopting the local coiffure.

some kind — no doubt a temple well known to archaeologists. It is situated about a hundred yards from the present-day coast of Lake Titicaca.

Frédéric Dumas, who is a specialist in marine archaeology, discovered in this same area a stone block on which a large serpent is carved. This reptile seemed to me to be a *naja* — a tropical snake not found at this altitude or in this climate. This may be additional evidence in support of the hypothesis that the first inhabitants of the Andes came, not from the Pacific coast of the continent, but from the Amazon basin.

A Llama Civilization

If we are so enthusiastic about searching for relics, and especially for pieces of ceramic, it is because we know that pottery has been found on the banks of Lake Titicaca dating from the third millenium before Christ. The carbon-14 test has left no doubt on that score — which means that 3,300 years before Christ, Lake Titicaca was the center of a great civilization and of a great religious faith.

In this civilization, the llama was both a beast of burden and a sacred animal. It was never used for food, however, for it was forbidden to kill a llama. The only exception was a sacrificial white llama, which was killed by the ruler as an offering to the sun.

Pre-Incan ruins on the shores of Lake Titicaca.

Frédéric Dumas and Professor Edgerton take a reading with a metal-detector.

Both the llama and the alpaca — the latter being a source of wool — are extremely valuable domestic animals, capable of living at high altitudes. They made it possible for the pre-Incan peoples to survive in an otherwise hostile country.

We saw many of these animals around Lake Titicaca, with their luminous eyes and their small pointed ears which are in constant motion. The natives attach pompoms of various colors to the ears in order to protect the animals from disease and accidents.

We have often admired the proud, stately carriage of the llama, as well as the strength which allows them to live independently of man in this rarefied atmosphere.

The Gate of the Sun

The religious center of the area around Lake Titicaca, and its most famous archaeological site, is the plain of Tiahuanaco, which seems to have been less the capital of a great empire than the location of a cult which was widely imitated from the Andes to the shores of the Pacific.

We visited Tiahuanaco to admire what remains of the temple, most of which has been transported to La Paz. There remains at Tiahuanaco the famous Gate of the Sun — which we found rather disappointing because of its small size. On the doorway, however, there is a relief of the god Viracocha, who dominates the whole history of Lake Titicaca. His head is surrounded by rays of the sun; and in each hand he holds a scepter ornamented with the heads of several condors.

The ruins of Tiahuanaco are spread over a large area. The dominant features are a large mound and a broken pyramid. The whole constituted a religious capital about which we know very little. It covered several square miles of land, and it is possible that its secrets lie buried today under the mud of Lake Titicaca. If we have one regret about our expedition to this region, it is that we were unable to dig deep enough into the mud to find anything that would help us to understand that mysterious pre-Incan civilization.

It is not that we gave up easily. Bernard Delemotte, in diving equipment, had Albert Falco tow him with the Zodiac for miles beneath the surface, looking for the slightest trace of artifacts. He finally had to call a halt when he developed vertigo. All that he ever found during these explorations were stones which stood out in the mud; but in no instance did they seem to be the remnants of buildings.

Jean-Clair Riant explored the area of the lake bottom where totoras grow. The Zodiac had a great deal of difficulty in maneuvering among these rushes, which became so entangled in the propeller that the motor stopped. Most of our explorations, however, centered around the bottom in the neighborhood of the Island of the Moon, on which, in the time of the Incas, the temple of the vestal virgins was located.

None of these investigations led to any discoveries, but they did succeed in awakening the interest of an "old man of the lake" in our comings and goings. This gentleman was born near the Island of the Sun eighty-six years ago. He informed us, as he chewed on his cola leaf, that there was a tunnel connecting the lake with Cuzco, the ancient capital of the Incas. This would make the tunnel some two hundred miles long. We followed up this lead — as we always did — but found no evidence of a tunnel of any size. But we did find, on this occasion, a few fragments of very ordinary pottery with no images painted on them.

Every day the air became colder; and every day clouds hovered among the mountains and sometimes sank down to the level of Lake Titicaca. It was November. The rainy season was beginning and would quickly put an end to our expedition.

It was no easier to load our equipment than it had been to unload it. Nonetheless, we managed to get everything to Puno, to load it onto the same incredible train that had taken us to the shores of Lake Titicaca weeks before, and to cross the Andes at the same breakneck speed.

As booty, we had in our luggage some items of geographic information, especially on the depth of Lake Titicaca, and a very vivid memory of that bottom covered with mud and green algae. We also had some specimens of frogs from the lake, which were intended for the Museum of Monaco, and the totora boat which we had bought.

By November 20, we were all back aboard *Calypso*, ready for a new adventure.

PART THREE
The Blue Holes

Calypso, moored over the Blue Hole of British Honduras.

CHAPTER EIGHT

The Sunken Cloister

André Laban had often spoken to me of the "Blue Holes" among the coral massifs of British Honduras and the Bahamas. A friend of his, Jacques Mayol, a diver who lives in Miami and is an expert on apnoea (he has dived to a depth of about 350 feet), has given him a detailed description of this strange geological phenomenon. These "holes" are caverns, originally hollowed out by fresh water and later filled with sea water. Several American magazines had run articles on these holes, and the photographs accompanying the articles showed some very interesting marine views. Laban finally convinced me that the Blue Holes were both a proper area for scientific investigation and a good subject for a film.

Laban was also acquainted with Dr. Benjamin, a chemical engineer from Toronto, who devotes his leisure time to exploring the limestone mazes dug out of the reefs of the Bahamas.

I asked André Laban to undertake a preliminary exploration of the area and to participate in the organization of our expedition by preparing for *Calypso*'s arrival and contracting for the supplies she would need.

Calypso was then en route from the Galápagos archipelago via the Panama Canal. The crossing of the Canal is very impressive, for the countryside is unspoiled and wild — especially when, at the highest point of the canal, one sees the impenetrable jungle growing up to the shore on both sides.

For a relatively small vessel such as *Calypso*, the navigation of the Panama Canal is a delicate operation, particularly when it comes to passing

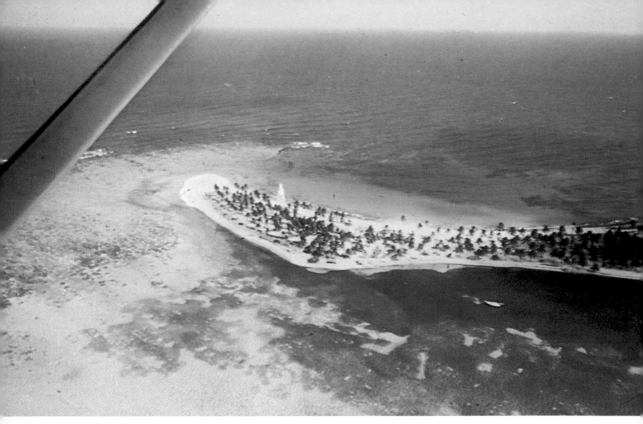

(Above) Aerial view of a part of Lighthouse Reef.

(Right) A careful watch is maintained at *Calypso*'s stem for dangerous coral protrusions.

(Below) The exterior edge of the coral plateau through which *Calypso* maneuvered to reach the Blue Hole.

through the locks, because of the strong currents.

Our first stop was at Belize, the capital of British Honduras. The countryside there is flat and marshy, and the climate is not the healthiest in the world, nor the most comfortable, being consistently hot and humid. The coastline had just been struck by a hurricane, and we could see the damage: trees destroyed, roofs blown away. It looked as though there had been a war.

The waters of British Honduras contain the second largest barrier reef in the world (the largest being the Great Barrier Reef off the northwest coast of Australia). Very deep water — up to 4,000 feet — separates this fringing reef from the mainland.

It was my intention to take *Calypso* to Lighthouse Reef, a coral reef situated fifty miles off Belize, where, Laban and his friends maintained, the largest and most interesting of the Blue Holes was located. In order to reach this cave on a direct course, however, *Calypso* would have had to pick her way across a plateau of coral which seemed almost at surface level. For navigation of this sort, we depend chiefly upon our eyes, for the charts of this area are not calculated to inspire confidence. The entire coastline of British Honduras is lined with a shallow level reef. The approaches to Belize are encumbered with the so-called "sunken keys"; and between the capital city and Lighthouse Reef there is Turneffe Island, formed of coral and covered with mangrove trees and lagoons. This coral topography is extremely complex and confusing; and, for that reason, we decided to take *Calypso* on a preliminary reconnaissance of the area.

Captain Bassaget was able to get her up to the edge of the cliff which is the outer boundary of Lighthouse Reef. But this island is nothing more than a mass of coral surrounded by other coral formations on all sides. I did not want to take *Calypso* any nearer until we had made a detailed survey with our launches, Zodiacs, and divers. On the south side of Lighthouse Reef, however, *Calypso* was able to go in close enough almost to touch the reef; for on that side there is an almost vertical drop into the sea which, at that point, attains a depth of 10,000 feet.

On our first dive, we discovered that the edges of the reef are decorated with a great number of large sea fans, sponges of all colors, and branched staghorn coral. Beyond this cliff, shimmering in the blue water, was the coral plateau, which seemed to be at a level with the surface of the water. Only our Zodiacs, which have a very shallow draught, would be able to be used — and then only at very low speeds and with many precautions. As for *Calypso*, she would have to remain at anchor alongside the reef, at least for the moment.

Even from *Calypso*'s observation platform, and with binoculars, we could not see the Blue Hole, seven miles away. In the Zodiacs, at water level, it was obviously impossible. We had to search more or less blindly, following the zigzag openings through the coral maze. Often the water was so shallow that even the Zodiacs could not get through; and then we had to get out and walk, pulling our boats along behind us.

We spent hours and hours exploring the reef in this way. Then suddenly, we saw it: a blue line in the distance.

We succeeded in reaching the Blue Hole the same day. It was an enormous ring of coral, containing a lake of blue-black water. Two of our divers could not resist it and went down immediately.

When it was time to return to *Calypso*, we could no longer see her. We had come so far that our ship was now below the horizon.

"It was an interesting search," said Yves Omer. Omer uses the word "interesting" only when things look really bad for us. It is his formula for expressing extreme pessimism.

The submerged cave known as the Blue Hole resembles one of the "rooms" of the Padirac pit in France or of Carlsbad Caverns in the United States, except that it has no ceiling. To my mind, it fulfilled all the glowing promises made by André Laban, and I was more eager than ever somehow to move *Calypso*, and all her equipment, to the site of the Blue Hole.

To do that, however, I would have to take *Calypso* across a coral plateau to which no ship's captain in his right mind would think of committing his vessel — almost eight miles, through shallow and treacherous waters, in the middle of a dangerous reef.

The solution seemed to be to find an open path through the coral. There was such a channel; but the divers reported that in places it was only three feet deep and about fifteen feet wide. Obviously, we were going to have a job on our hands in our efforts to find a way for *Calypso* to reach the Blue Hole. Yet, the fact that it would be so difficult to reach our objective made it seem all the more fascinating to us.

I sent out divers to locate the largest massifs of the reef, hoping that around those massifs they would also find a channel sufficiently deep for *Calypso* to pass; but they returned to report that there seemed to be no such channels, and that the reef was often barely below the surface of the water. Coral formations — the kind known as Neptune's brain, because of their peculiar shape — protruded menacingly everywhere on the surface.

Still, I had the most complete confidence in *Calypso*'s ability somehow to

(Following page) The Blue Hole is an almost perfect circle of coral. There was only one break through which *Calypso* could pass.

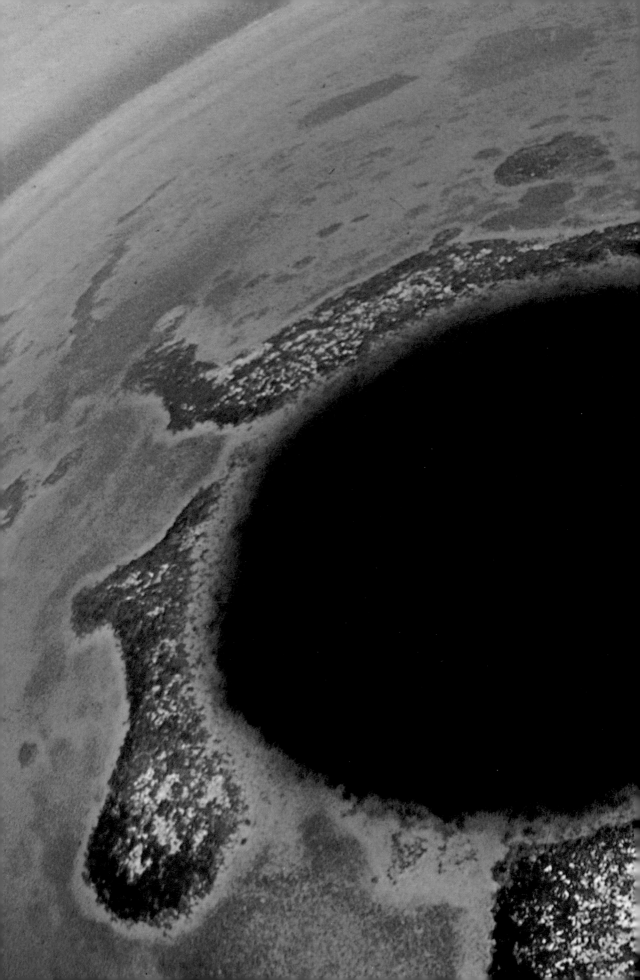

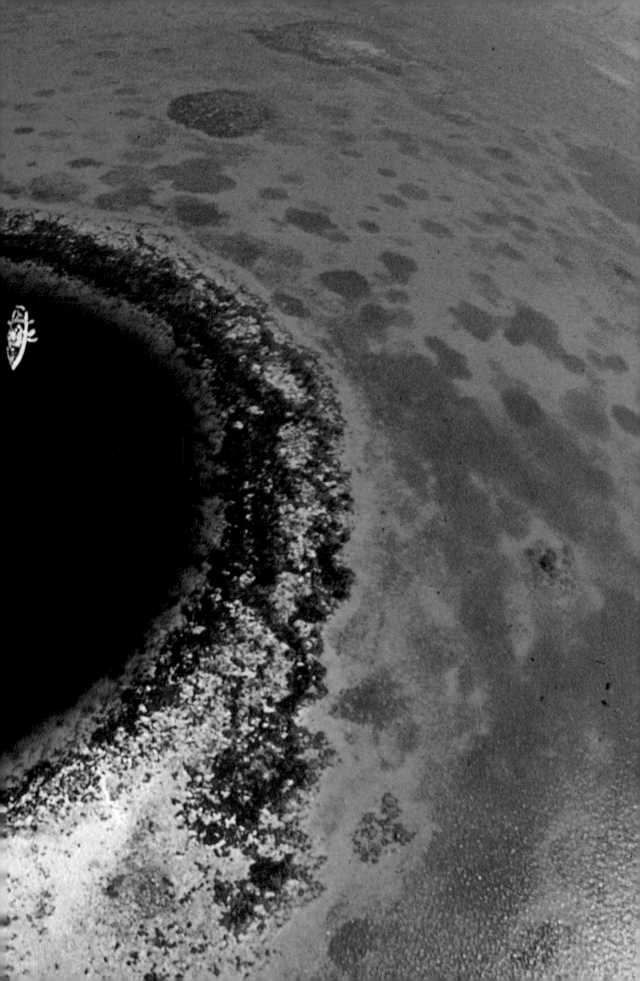

reach the Blue Hole. She had been in impossible situations before; and in the Red Sea, she had performed incredible feats of maneuvering among the reefs of the Far San Islands and the Suakin archipelago.

I was perfectly willing to take necessary risks, but I was unwilling to take unnecessary ones. I therefore decided to put ashore a team of divers, who would be responsible for finding the least dangerous route through the reef for *Calypso*. The rest of us would take the ship to New Orleans for repairs on the false nose and the observation chamber, which had been damaged as the result of our encounter with an uncharted rock in the Galápagos archipelago.

Calypso's damages were considerable. The hull was untouched, and there were no leaks. The false nose had served as a bumper and had saved us from more serious damage. But, as a result, it, and the observation chamber, were crushed. The observation chamber was especially important. It is beneath the water line, and it is a valuable aid in such undertakings as picking one's way through a coral reef. An observer stationed there, and in communication with the bridge, can save us from a disaster.

A Hazardous Passage

We remained in New Orleans for a month, and when we returned to Lighthouse Reef, *Calypso* had a new false nose and observation chamber. We dropped anchor in the same place as before, alongside the south side of the reef.

During our absence, our divers had found and marked a channel through the coral plateau, leading to the Blue Hole. Even this route, however, was going to be very difficult to navigate. *Calypso* would be obliged to make some sharp turns. And, despite our team's best efforts to find the most navigable passage possible, there were places where *Calypso*'s keel would be practically on the bottom.

It was worth trying, but I was still unwilling to run risks that were not absolutely necessary. I wanted to be certain of the weather, and to take advantage of the tides — which, unfortunately, are quite weak in this area.

As a further measure of security, I had hired a helicopter to hover above us and guide us by radio through the channel laid out by our divers. In addition, we installed a portable sounding device in one of our launches, which, theoretically at least, would keep us in no less than twelve feet of water. We made a trial run in the launch, using *Calypso*'s radar to keep track of its position as it set out the buoys which would serve as our guides through the channel.

We had now done all that we could to insure a safe passage for *Calypso*. Nothing that we did guaranteed success, of course, but we had every hope that, by using the utmost care, we could reach our destination safely. Even so, we had no illusions. We remembered vividly our navigation of the Silver Bank during our search for the treasure ship *Nuestra Señora de la Concepcion**, and we knew that we were in for the same sort of hair-raising ordeal.

Albert Falco was in charge of positioning the plastic buoys along *Calypso*'s route. He used yellow ones on the left side and red ones on the right. He also organized a team of divers to precede us through the canal and keep us informed of the shape and size of any obstacles in the ten miles between *Calypso*'s anchorage and the Blue Hole. All told, Falco's team set out thirty-two buoys. In addition, he put luminous buoys at the most dangerous places in the canal, in case of a sudden storm at night, while we were within the reef.

Two bridge officers, Yves de Pimodan and Laval, had charge of the chart showing the route *Calypso* was to follow. On this chart, they entered the location of every buoy.

All possible precautions having been taken, I began discussing the details of the passage with our captain, Jean-Paul Bassaget. We agreed that it would not be possible for us to avoid all risks; but we also agreed that the obstacles *Calypso* might encounter must all be studied and taken into account so as to reduce the danger to a minimum.

Finally, we were ready. *Calypso*, preceded by two launches, moved onto the coral plateau. A launch was sent out with several divers, including Yves Omer, to join the Zodiacs. They would make periodic dives to inspect the canal's bottom. There was also a man on duty in the observation chamber, studying the canal from his position eight feet below *Calypso*'s water line. He reported that the bottom was sandy, but that the walls of the passage alongside the hull, on both sides, were studded with coral formations. So far, so good. But our friend in the observation chamber was obviously worried about the shallowness of the water in the canal. Being an old hand aboard *Calypso*, he added that he could see actual mounds of mollusk shells — conchs and spider conchs. This area is famous for its shells, and fishermen often come to the reef to gather them.

Yves Omer was now in the water watching *Calypso*'s slow progress. He saw the stem directly in front of him — and it seemed to be scraping bottom. Later, he told us: "It was very impressive. I knew that *Calypso* was moving slowly, but from beneath the surface, and in such shallow water, it looked as

(Above) The level reef of Lighthouse Reef breaks the surface almost as far as we can see.

(Upper right) The shallow area surrounding the Blue Hole is covered with sponges, coral and sea fans.

(Lower right) Our cameramen begin their exploration of the underwater cave.

though she was moving at full speed — and coming right for me. I could see her propellers turning. It was a strange feeling."

By the time we were two miles inside the reef, I had the curious sensation of having left the sea, of having entered a new world; it was as though I were trapped, and *Calypso* had entered a maze from which it would be impossible to escape. All around us there were coral massifs, many of which rose above the surface of the water. On the bridge, with one hand on the sounder and the other on the wheel, I could not even believe that we were in a channel. It looked more as though we were simply adrift in an endless sea of coral.

As we went forward, the water became more and more shallow. There were eight feet of water under our keel; then, suddenly, there were five feet. Our sonar showed places where there were only three feet. Fortunately,

Calypso is equipped with a device — a hand accelerator — which makes it possible for me to control her speed directly from the bridge. It proved extremely useful at this time.

There came a moment when the divers and launches reported that we were coming into an area in which coral formations protruded from the bottom which, until then, had been only sand. I decided that it was no longer safe to leave a man in the observation chamber; for the chamber might, at any moment, be crushed against the coral. Instead of a human observer, therefore, we installed a television camera in the chamber, which relayed a picture of the bottom to the bridge. And then a nightmarish announcement came through our walkie-talkie: "*Calypso*! *Calypso*! Only five feet of water at azimuth 18.5!"

Calypso Goes Aground

Ahead of us, there was a section of channel which was passable only if we were capable of the most adroit maneuvering. A yellow buoy marked the spot where *Calypso* would have to make a sharp double turn in order to avoid a coral massif at surface level.

Jean-Paul Bassaget acquitted himself nobly — but *Calypso* went aground. We had done everything humanly possible to avoid such an accident, but it had happened nonetheless. However, it was not as bad as it could have been. *Calypso*'s wooden hull had not run onto the hard, cutting coral, but merely settled gently into a stretch of sand.

Divers were in the water at once, inspecting the hull for possible damage. They reported that the only visible effect had been some scratches in *Calypso*'s paint. It was a great relief. But the fact remained that, damage or not, we were aground, and we would have to extricate ourselves somehow.

So far as I could see, there was only one way to do so. I ordered the aft auxiliary anchor — which works on a windlass — to be lowered. Then I had the launches move to a position alongside *Calypso* and begin pulling her sideways, so that she would — I hoped — slide down the bank of sand into somewhat deeper water. At the same time, we threw our engines into reverse. It worked. With the anchor holding her stern steady, the launches pulling, and the engines racing in reverse, *Calypso* was suddenly free again. Once more, we began moving forward slowly — and, needless to say, with redoubled caution.

The feeling that we got in moving across that coral plateau was very strange. From the bridge, it looked as if we were moving across a partially submerged island. For a sailor, it was a situation out of a nightmare. And it did not help that we knew that at any moment *Calypso*'s hull might be split open by one of the countless coral heads which protruded along both sides of the channel.

We had reached a position between the sixth and seventh buoys, and we began scanning the reef for a glimpse of our destination. But we had not yet gone far enough. There was no sign of the Blue Hole.

Finally, we passed the last buoy, and ahead of us we could see a stretch of dark water. The indigo water of the Blue Hole formed a lake in the middle of the coral and the lighter blue of the sea.

The diameter of the Blue Hole is approximately a thousand feet, and the hole itself is neatly circumscribed by an almost perfectly circular belt of coral, broken by two narrow passages, which protrudes above the surface of the water. There is a lip of sand around the interior of this belt — an indication of the steepness of the interior walls of the hole. At that time, I estimated the depth of the Blue Hole at about 350 feet. Legend says that it is "bottomless." Legend also says that it is inhabited by sea monsters and that it swallows up any vessel which dares approach it.

Having passed the last buoy, we were in the waters of the Blue Hole itself, and we had as much depth as we needed. We had expected a great deal of *Calypso*, and we had not been disappointed. She had done what no other vessel had ever done.

Calypso's men went to great trouble to prepare a secure mooring for her, wrapping steel cables around great coral massifs, through which we passed our hawsers. The ship was secured in a position convenient for our divers and also for unloading the equipment which we would need in our exploration. This equipment consisted chiefly of our two minisubs, a special launch for moving the minisubs, and our new wet submarine. This latter vehicle is most useful for its speed under the surface and for its cargo of air bottles and breathing equipment — a load which our divers no longer had to carry strapped to their backs. Three newly modified underwater scooters are also used to carry such equipment.

Aboard *Calypso*, we left the Chaparral, a new six-wheeled amphibious vehicle, which we intended to use for the first time in the Bahamas later on during the expedition.

The First Dives

In the days following our arrival at the Blue Hole, André Laban, Albert Falco, and Yves Omer began diving into the dark blue waters of the hole. We

(Above) Divers inspect the stalactites of our "underwater cloister."

(Left) Bob Dill takes a geological sampling from the vault of the cave.

did not know its topography, but we sensed that it was very deep. Our regular divers were fortunate enough to have the company of a friend of André Laban, an American named Bob Dill. I wanted to have a scientific adviser with us during our explorations of the Blue Holes, and I had asked Bob, who is attached to the U. S. Navy's Undersea Research and Development Center, to join our expedition. We expected to make much use of his expertise in marine geology — especially when it came to the complicated formations of these Honduran waters.

The coral ring which belted the Blue Hole was itself very blue. Large sponges and sea fans made it a living border, and a beautiful one. Yet, in the back of my mind there was a nagging thought. What if there were a sudden storm? How would we ever get *Calypso* safely across that ring? I could not help remembering all the tales we had heard of ships which had entered the Blue Hole never to be heard from again. It was not hard to believe that there was some basis in fact for those legends.

The wall along which our divers descended was riddled with holes. The water was clear compared to that of the coral plateau, where there is always sand in suspension. We found that the Blue Hole was inhabited — by sharks, tuna, and barracuda. It must be just as difficult for them to reach the hole as it had been for us. There are only a few means of access through the coral plain.

There seemed to be very few coral fish in the Blue Hole. But we amused ourselves with some rays and inoffensive sand sharks, which we chased briefly.

The walls of the Blue Hole curve outward at a depth of about a hundred feet, and the pit becomes wider at that point. This seems to be an indication that, at some point in time, the cave had a roof or cupola which has since collapsed.

Pieces of limestone were continually coming loose from the edge of the hole and sinking gracefully downward, like snow.

At 125 feet, on an overhang, the divers discovered a massive formation of vaults, hanging columns, and of side "chapels." In the light of their lamps, it resembled a Gothic cloister. Some of the stalactites were forty feet in length. Several were over six feet in diameter. And a few were twisted like the pillars of a cathedral. But only those in the lowest part of the vault touched both the floor and the ceiling of the cave and were true columns.

The first impression in this forest of stone was one of size. The films and photographs taken of this marine wonder are true to life, but they do not convey adequately this impression. These gigantic pillars and columns must be seen in the relatively small space which they occupy if one is to receive the impression of gigantism which we did. Even so, the size of the pillars themselves comes across clearly in our photographs and films.

After spending some time in this unearthly cloister, we returned to the surface. Bob Dill, our geologist, was greatly excited. Without taking the time to change from his rather bizarre diving outfit — red shirt, flowered shorts, and black socks — he began describing what he had seen. He could hardly control his enthusiasm. And it was no wonder, for he had seen what no man had ever seen before. The formations are stalactites — but they are *underwater* stalactites. The significance of this is that this Blue Hole was once a true cave — and that it was *above* the surface of the water. For stalactites can be formed only in the open air. The origins of the Blue Hole can therefore be traced to the end of the glacier era, when the sea rose and filled the cave. The ceiling of the cave then collapsed — forming the Blue Hole.

The next day, we launched the two minisubs. We wanted to know what there was at depths to which divers could not go. André Laban piloted one minisub, and Albert Falco the other.

At 150 feet, Falco saw a long ledge along the side of the hole and inspected it thoroughly. It was almost horizontal and about three feet deep.

As soon as Falco notified *Calypso* of this discovery, Bob Dill went down for a look at it. He reported that it was a "fossil path" — which meant, he explained, that it marked the level of the water at some time in the past. This was about twenty-five feet below the level of the cave which we had explored the preceding day.

Meanwhile, the two minisubs were continuing their descent into the hole. At 250 feet, Laban observed that the water, which had been clear until then, was becoming cloudy. That was probably due to some sort of bacterial fermentation. Also, at this depth there were no more fish. The water was very likely sulfurous, for parts of the minisubs were stained, and later we found that the water in some of the vehicles' equipment had the smell of hydrogen sulfide. This proves that there is no current, and no exchange between the bottom of the Blue Hole and the ocean. The water of the hole is motionless, despite the tides — which are quite weak. Certainly, at one time there must have been an opening between the sea and the hole, but it has since been blocked — possibly by the collapse of the roof, or even by the action of the sand.

At 400 feet, the two minisubs reached the bottom of the Blue Hole. Laban and Falco noticed that there were large pieces of broken rock strewn about — no doubt pieces of the roof — and a very old layer of mud and limestone debris, which they were careful not to disturb since this would have made the water dangerously cloudy.

As the two miniature submarines crossed the lunarlike plain of the hole's bottom, Falco caught sight of another overhang and some small stalactites. This was an important discovery, for it indicated that the bottom of the Blue Hole, at some point in geologic time, was above the surface of the sea. That is, that the water level was at least 400 feet lower than it is at present.

Falco and Laban rose slowly back to the surface. They had made discoveries of scientific significance during their dive; but they had failed to discover the monsters described in the many legends concerning the Blue Hole.

An Accident

I wanted to film all the details of the unusual geological formations in the Blue Hole and especially the stalactites in our "Gothic cloister." In order to get good shots, however, we had to have powerful electric floods. The installation of such is always difficult and dangerous, since we have to run electric cables from *Calypso* to our filming site. On such occasions, we recruit every hand aboard *Calypso*, including our ship's doctor — in this instance, Dr. Blanc. It is a hard job. The cables are very heavy and require a great effort to move. For someone who is not accustomed to it, it is dangerous to expend much energy in deep water; and Dr. Blanc was not accustomed to it. At 150 fcct, he lost consciousness. Fortunately, Yves Omer was nearby and saw immediately what had happened. "The doctor was holding some cables in his hands, and we could not get them away from him. We had to take him back to the surface, dragging the heavy cables along with us. It was quite a weight; and of course we had to make a decompression stop. I could see his eyes beneath his mask; but I could see only the white, and no pupil. We tried to revive him by slapping him lightly, and then we stretched him out on a sand-covered ledge, where he regained consciousness. Then we took him to the surface."

Finally, Dr. Blanc was pulled aboard *Calypso*. He was quite certain that the divers had saved his life, and he was probably correct. At 150 feet, he had lost consciousness entirely and was motionless. Under the weight of the cables he was holding, he would probably have begun to sink if Yves had not been at hand.

The incident reminded Yves of a similar experience he had had during our third sojourn at Conshelf III, our "underwater house." The divers were dismantling the house at the end of the experiment, and Yves was carrying some electrical cable to the surface. Yves had to climb up a slope to reach the surface, and the cables were so heavy that, every time he started climbing, he would fall and slide back to the bottom. On that occasion, Albert Falco was nearby in the minisub. He picked up Yves and took him back to the Underwater House.

The entire maneuver presupposed great presence of mind on Falco's part, sufficient intuition to sense that Yves was in serious trouble, and above all a great dexterity in order to be able to use the minisub as effectively and gently as he did.

A Born Actor

Bob Dill also gave us a couple of scares. Bob is not only an excellent diver, but a movie buff, and he loved to be in the underwater scenes that we

The stalactites are covered with fixed animals.

were filming. He was more than willing to do whatever he was told to do by the cameramen, and he was very good at coming up with ideas for new sequences. He always managed to place himself in precisely the right position for a shot. He was an ideal actor for our purposes.

On one occasion, he worked out a sequence in which there were to be three main elements: himself, his geologist's hammer, and a stalactite. "I'll go down," he told the cameramen, "and approach the stalactite from the left. You stay on the right side of the stalactite, and you will be able to get shots of me removing a piece of the stalactite with one stroke, putting it in my bag, and then rising to the surface."

The cameramen got into position and watched Bob coming down. He approached the stalactite, raised his hammer, and then brought it down — on his finger.

Aboard *Calypso*, Dr. Blanc took three stitches and bandaged the actor's finger.

A short time afterward, Bob had an encounter with a stubborn remora. These fish usually attach themselves to sharks by means of the oval sucking disk at the tops of their heads (they are also known as shark suckers). This particular remora, however, found a home on Bob's shoulder. It required much ingenuity on our part to pry the remora loose without letting it take Bob's shoulder with it.

CHAPTER NINE

The Secrets of a Stalactite

On board, we were very conscious of the situation in which *Calypso* would find herself if there should be a sudden storm or a hurricane. Even if there were only a swell, it would have been virtually impossible for us to be able to retrace our course across the coral plain to the open sea. In other words, we were in the same predicament as we had been when we were diving down to *Nuestra Señora de la Concepción* in the Silver Bank.

It is true that, basing ourselves upon our experiences in the Silver Bank, we were able to place marking buoys in the channel leading through the coral plain to the Blue Hole. But I kept wondering whether I would dare commit *Calypso* to that same precarious channel in bad weather.

Claude Caillart had been our captain during our treasure hunt on the Silver Bank; and, on that occasion, the official weather bulletins had all announced that a hurricane was heading our way. But Caillart, relying upon instinct, had assured us that the hurricane would pass farther to the north — and he had been perfectly right. I kept hoping that we would have as much luck at the Blue Hole of British Honduras.

To be on the safe side, I decided that we would have to do more work on our mooring, to assure that we were as secure as possible. A team spent a

(Following page) The whole of the marine landscape along the coast of British Honduras is dominated by coral reefs inhabited by brightly colored fauna.

whole day checking the seven points to which we had run lines. The divers selected the very largest coral massifs in the immediate area and tied steel slings around them. Then we once more attached our nylon hawsers to these cables.

I took advantage of this occasion to change our position slightly. We were now tied to two solid massifs in front of *Calypso*, three across and two behind. All we had to do to move *Calypso* directly over the hole was to haul her to the rear; and, by the same token, we could just as quickly move her away from the edge of the hole and resume our original position.

The submerged cloister in which we had discovered the stalactites faced west, and *Calypso* was now positioned above its two entrances. One of these is narrow, and the other wide. Both are at about 150 feet.

Everything was now ready for a systematic exploration of this unique natural formation.

In the days which followed, our divers, under Bob Dill's supervision, undertook a very detailed scrutiny of the Blue Hole. Bob, by his good humor and his ability as a diver, combined with his skill as a marine geologist, had become everyone's favorite. He had been quite firm in his refusal to use one of our diving suits, and he insisted on wearing his own picturesque outfit — his red shirt, flowered Bermuda shorts, and black slippers. His long nose, which was always crushed against the mask, and his protruding ears, gave him a rather comic appearance.

Bob was convinced that the mysterious Blue Hole of Lighthouse Reef was of great scientific interest, and he believed that our dives would make an important contribution to man's knowledge of the history of the earth.

"It is in natural pits like this one," he said, "that we are best able to study the changes in the level of the oceans during the last glacier era, when the water sank in all the seas of the world as the result of the invasion of the continents by ice. Of course, this lowering of the sea level did not take place all at once, and the Blue Hole certainly contains indications of its successive stages."

Bob Dill's faith was contagious. The Blue Hole now seemed a treasure chest of scientific truth, and we brought a new enthusiasm to explorations which confirmed what we had learned during our initial dives.

Near the surface of the water, the Blue Hole begins with a vertical wall 10 to 12 feet high. This wall is completely covered by a luxuriant growth of coral and sea fans, extending around the entire circumference of the hole's opening — except for two areas, each one about 65 feet wide, on the north and east sides.

Starting at a depth of 12 feet, there is a slope of 15° which meets the vertical wall at 50 feet on the north side and at nearly 60 feet on the south side. Bob Dill immediately saw that this difference between the two walls was of capital importance, and, on the basis of it, he concluded that there was a shift in the hole at a very remote era.

In this upper part of the Blue Hole, there were many coral formations, some of them 10 feet long, mixed in with sponges, coral algae, and sea fans. Fish grazed on the coral, which they convert into sand; they regurgitate the sand, which drifts down into small crevices.

The vertical wall which began at 50 feet was much less rich in fixed animals. Even so, it was covered with patches of calcareous algae and with a thin sprinkling of sea fans. This wall goes down to a depth of about 100 feet. Then the hole becomes wider and forms a vast cavern, the ceiling of which slopes at a 55° angle.

Horizontal ledges and small caves were found at two levels: 70 feet and 160 feet.

All these data served to confirm what we had already discovered: that the Blue Hole was of karstic origin and had been formed at a time when the water level was more than 400 feet lower than at present; that is, beneath the maximum depth of the Blue Hole. In more recent times, the rising water had inundated the reef; but the horizontal ledges which we found at 70, 160, and 300 feet marked the levels at which the water level had remained stationary for a time. (Similar indications are also found in the limestone depressions of Yucatan, which are known as *cenotes*.)

Leaning Stalactites

The cave with stalactites is located at a depth of 160 feet. In company with Bob Dill, we examined it thoroughly. Bob quickly noticed that not all of the stalactites are vertical and that even the angles of individual stalactites vary from 10 to 13 degrees. Their slant is in a southerly direction. Near the walls of the cave, however, most of the stalactites are vertical.

Our divers could not help smiling as they watched Bob Dill rush from one column to the next, frantically measuring the angle of each one with a small instrument which he had made, and scribbling feverishly on a rhodoid slate.

About 10 feet farther down, Laban made a discovery. He saw a large stalactite lying on the path or ledge which runs all around the cave. Raymond Coll, who was with him, rose to the level of the cave's ceiling, about 30 feet above, and found the place where the stalactite had been formed and from which it had fallen.

(Above) Bob Dill illuminates a wall of the underwater cave.

(Left) Our divers go down to attach a sling to the stalactite we have decided to hoist aboard *Calypso*.

At the end of this dive, we had an excited exchange of opinions with Bob Dill. According to what he had seen, he estimated that in the interim between the formation of the slanted stalactites and that of the other stalactites, there was an upheaval of some kind which resulted in the slanting of the entire plateau known today as Lighthouse Reef. It is possible that, during this upheaval, the vault of the cave collapsed and that some of the stalactites were broken. Over all, it seems that this entire area was affected by some great cataclysm at a point in time which Bob wished to pinpoint.

It is possible that the stalactites were formed during one of the glacial regressions of the Quartenary Era. According to our observations, a first series of stalactites shifted from 10 to 13 degrees southward; while a second series — those which remained vertical — show that the reef has remained stable at least from the time when the level of the sea was at least 175 feet lower than it is now.

The Isle of the Damned

We gradually became familiar, not only with the Blue Hole over which we were anchored, but also with the unusual topography of the reef itself.

Fishermen often came to those areas of the reef where fish were plentiful, especially around and over the keys. From what we could see, these fishermen were not a healthy lot, and our doctor distributed as much medicine as he could among them and tried to attend to their needs. On some days, he had as many as twenty patients from among them.

The fishermen told us that around the key to the north we would find alligators. I was very interested in this information, since I was working on a film about marine reptiles. I therefore sent Jean-Paul Bassaget with a team to see what they could find. They came back at the end of the day, exhausted. The island to which the fishermen had referred was difficult to reach, and it was not very inspiring: a soft, spongy soil enclosing a lagoon filled with stagnant water. The stench of the place, they told us, was unbearable. All of the trees were dead, and their stumps were filled with mosquitoes and flies — all of which fed hungrily on our men. Apparently, they had never had such a feast.

The environs of this virtually uninhabitable island turned out to be filled with life. Jean-Paul and his companions saw sea turtles, sharks, rays — but no alligators. On the other hand, they did see some black-and-white iguanas (land iguanas, and not marine iguanas such as live in the Galápagos) and a large number of birds which had apparently never been disturbed by human beings: pelicans, herons, yellow-beaked waders, and sea eagles.

The Whims of a Miniature Submarine

On May 21, at 1:30 in the afternoon, Falco left for a dive in one of the minisubs. At 250 feet, he discovered a crevice in one of the walls of the Blue Hole, but he was unable to explore it, for the water was cloudy. Only ten feet lower, however, it was quite clear.

Falco's gyroscope was not working; and all sorts of debris falling from the edge of the hole was striking the hull of the minisub. He did not dare take the chance of entering the crevice — an act of prudence of which I approved

Jacques Renoir, from a precarious perch on *Calypso*'s crane, prepared to film the raising of the stalactite.

wholeheartedly. So, he contented himself with taking pictures of the small jellyfish and shrimps circling around the minisub. Then, leaving the wall of the pit, he went toward the center of it where we had sunk a nylon line which was suspended from a buoy on the surface. The line ran down to the bottom of the hole, and Falco followed it downward to a depth of 325 feet, but saw nothing remarkable.

The following day, May 22, was one of great activity. During the morn-

(Above) The stalactite rises to the surface surrounded by divers.

(Left) The stalactite is slowly lifted out of the cave in which it lay for thousands of years.

ing, we dived in groups and shot films to a depth of 200 feet. While we were working, I saw a handsome stalactite on one of the fossil paths. Apparently, after becoming detached from the ceiling, it had landed on this ledge. It seemed to me that we might be able to tie a line around it and hoist it aboard.

The afternoon was devoted to a new exploration in the minisubs. We wanted more information on the vault and on the crevice which Falco had found the preceding day. Falco therefore took one of the minisubs, and Raymond Coll took the other.

The two minisubs went down to a depth of 300 feet. Falco then decided to explore the cone of mud on the bottom and headed in a northerly direction. The cone, he discovered, was rounded, and its edge touched the sloping wall of the hole. The water around it was surprisingly clear, and Falco could see no reason why the minisub could not pass between the pile of mud and the wall.

He was between them when his engine went dead. The first thing Falco thought of was to release his lead ballast; but there was none left. He then shut off all systems and attempted to start up the engine. Nothing happened.

The extremely soft mud seemed to exercise a magnetic attraction on the minisub and began slowly to envelop it. Falco, understanding what was about to happen, now released his emergency ballast, but the minisub remained in the mud.

Meanwhile, Raymond Coll was about 300 feet away, trying to get his own minisub into the crevice which Falco had found earlier. From this position, he could not see what was happening on the other side of the mud cone. Otherwise, he would have been able to use his exterior pincer to pull Falco's minisub to safety. The minisubs were designed, in fact, to be able to help each other.

By now, Falco was losing hope. Then he had an idea. He began shaking the forward part of the minisub as violently as he could, hoping that the vibrations would loosen the mud and release his vehicle. It was successful. Moreover, he was able to start the engine, and the minisub began to rise.

But Falco was not out of danger yet. The minisub rose briefly, and then crashed into the sloping ceiling of the cave and began scraping along the rock, like a bird in a trap. Falco was able to free it, and the vehicle began moving again. But it was still against the wall, and Falco was able to see the fossil path ahead of him. In less than a minute, the minisub would collide with the path. At that point, however, Raymond Coll had been notified by telephone of what was happening. His minisub approached Falco's, took hold of it with its pincer, and pulled it out into the open water in the center of the hole. Then the two minisubs climbed toward the surface, side by side in the increasing light. The launch was there to meet them. Once more, Falco's cool head had saved him.

A One-ton Trophy

In order to be able to carry our observations as far as possible, I decided to bring aboard the stalactite I had seen lying on the fossil path some 150 feet beneath the surface. Bob Dill and I hope that this column may provide the answers to a few unanswered questions: when, precisely, did the variations in the water level occur? How long ago did the upheavals occur which affected this entire region?

It would not be an easy job. The stalactite was quite large and no doubt very heavy. Moreover, it seemed to be partially sunk into the sediment on the path.

We began by positioning *Calypso* over the column which we intended to hoist aboard, so that the hoist on her rear deck could reach the stalactite. To do this, however, we had to change our anchorage and move *Calypso* farther into the Blue Hole.

While we were doing this, a team went down to disengage the stalactite so as to be able to attach a line to it. It turned out that it was resting in dead coral — which the divers had to break away with pickaxes.

Since I wanted to film the entire operation, we had once more to install the electrical cables for our lights.

Once our underwater studio was ready, the column disengaged from the coral, and lines attached to it, we lowered a heavy cable from *Calypso* and hooked it to the lines. The hoist went to work, and the stalactite, in a great cloud of mud, began slowly to rise. On *Calypso*'s deck, our windlass strained under the pull of a trophy which weighed one ton and was twenty feet long.

Our divers remained close to the stalactite while it was being raised to the surface, preventing it from striking the walls of the Blue Hole and from becoming caught in any crevices in the walls. The operation took a long while, and our cameramen had the opportunity to film it from every possible angle.

Finally, the stalactite reached the surface and we could see it clearly. It was a magnificent sight. The marine fauna clinging to it gave it a variegated hue, except for one side — that which had been covered by sediment, which was snow white. It shone with a strange luster under the sun as it rose out of the dark blue water and the silver foam.

How to Cut a Stalactite

Once the hoist had set the column down on *Calypso*'s rear deck, we realized how large it actually was — and how much of our valuable space it occupied. Bob Dill must have noticed our rather stunned looks. He said, "Well, I never asked for the whole thing. All I wanted was a small piece."

The trouble was, we had no idea how to go about cutting a stalactite in such a way as to get a good slice of it. For these formations are like trees, in the sense that a slice of it exhibits the inside layers; and it is these layers which give geologists the information they need.

Unfortunately, even though *Calypso* carries tools of every sort, we had no saw capable of cutting through stone. There are some things which cannot be foreseen. . . .

Finally, we worked out something which was extemporized but effective. At the end of a slender steel cable, we attached some heavy lead weights and then lowered them into the water by means of our windlass. We passed the cable over the stalactite in such a way that, as we raised and lowered the lead

(Left) The heavy stone column is finally out of the water.

(Right) Our geological trophy is to rest on *Calypso*'s rear deck.

(Below) A slice of the stalactite. It will be a mine of information for geologists on the history of the earth.

Bob Dill, surrounded by the divers who brought the stalactite to the surface, takes the measurements of his treasure.

weights, it would rub against the stalactite and, we hoped, cut into it. It worked; but it worked slowly, and we had to do a lot of raising and lowering before the cable had cut completely through the stone. The operation went on, without stopping, from mid-afternoon until two o'clock the following morning.

The chief problem with our stone-cutting cable device was that a great deal of heat was produced by the friction between the cable and the stone. Finally, we installed a pump on deck, and poured a steady stream of water over the whole thing, both to wash away the stone dust and to cool our ingenious saw.

In the middle of the night, once the sawing had been completed, Bob Dill began immediately inspecting the slice we had removed, no doubt expecting to be able to read in a glance the whole history of the sea. But the polished slice of stalactite, beautiful as it was with its concentric rings varying in color from brown to ocher to white, proved to be a geological document which was very difficult to decipher. He then began smelling it — and putting bits of it into his mouth to taste it. It seems that this is one of the classic methods by which geologists recognize certain kinds of rock.

A few days later, we found and raised another broken stalactite, which we had found on the same ledge as the first one. This one was not as large or as heavy, and it was easier to handle.

All told, our team spent about two months studying the Blue Hole of Lighthouse Reef, and I am of the opinion that we did everything possible during that rather long period to insure that our investigations would prove valuable from a scientific standpoint.

At the same time, however, we had a television film to shoot. I decided one morning that we could make a good sequence of our new scooters in the interior of the cave. We therefore sent out Yves Omer and Raymond Coll on the scooters, and they had a marvelous time zooming around the stalactites, diving down toward the bottom, and then rising almost vertically toward the surface. It was like a bizarre underwater ballet, with the two scooters crossing paths and then pulling away from one another in a stream of silvery bubbles, while, in the light of our floods, our cameramen filmed away untiringly.

A Spectacle of Devastation

I intended to continue exploring the coral complex which surrounded *Calypso*, for I believed that these reefs might hold a few surprises for us. Moreover, we were constantly being told that we would no doubt encounter alligators.

I went with a team to North Cay, where everyone says there are all-igators in abundance. The team's experiences were reminiscent of those of the team sent to North Key. With great difficulty, we crossed dismal swamps — in which we occasionally floundered — before reaching an inland lagoon of stagnant water. We found crabs, eagles, lizards, and iguanas — but no all-igators.

While we were gone, the divers aboard *Calypso* spent their time more enjoyably than we did. Everyone dived among the lichens in the Blue Hole. It was, I am told, absolutely marvelous.

I also organized a trek to a large island nearby, Turneffe Island, located between Lighthouse Reef and the Honduran coast. I thought that the island might be a welcome relief after so many weeks of nothing but coral reefs and swamps. I was wrong. What awaited us on Turneffe Island was a spectacle of devastation. The trees were all splintered. The mangroves lay flat. It was as though this island had been subjected to saturation bombing. It was covered with debris, and ruin was everywhere. It had been ripped from one end to the other by the most recent hurricane.

There were many stretches of swamp and many lagoons. The lagoons contained coral, which had also been damaged by the hurricane, and many tropical fishes of various colors.

For several days, we had bad weather, with the wind blowing from the northeast at thirty knots.

Our mission was now over, and I was eager to get *Calypso* to Belize, for we were low on supplies and especially on water. We were already on rations, and no one was allowed to wash in fresh water. However, I did not want to take *Calypso* across the coral plateau, through our narrow and shallow channel, in such weather. The waves were breaking against the side of the reef, and even over the reef. In the Blue Hole, with the entire width of the reef protecting us, we were in calm water. And, with 400 feet of water below us, we were sheltered. There was nothing to do but wait.

Finally the weather cleared up as quickly as it had degenerated, and we decided to take advantage of it immediately. *Calypso* left her anchorage and once more entered the channel through which she had reached the Blue Hole.

We took the same precautions as before. The Zodiacs and divers went ahead of us. I kept my eyes on the sounder and on the television screen which relayed a picture of the bottom from the observation chamber. We were worried when we reached the spot at which we had run aground during our first crossing; but now we knew how to handle ourselves and we passed safely.

It took us seven hours to reach Belize, but we took on fresh water immediately and everyone bathed and showered to their hearts' content.

Then we went into the city — a sleepy city wrapped in subtropical humidity — where we were delighted to visit friends who had been of much help to us. One of the most useful of these friends was a man named Johnny, a radio ham with whom we had been in constant touch since we reached Lighthouse Reef and whose patience and co-operation had greatly facilitated our work.

During our brief stay at Belize, several people told us that there were dugongs along the coast of British Honduras. Since we were planning a film on the large marine mammals, I would have been delighted to find some of these animals, and once more I sent out Jean-Paul Bassaget and Michel Deloire to look for them. Unhappily, they found none. I was told later that the last time dugongs were seen in the waters of British Honduras had been a dozen years before.

From the Blue Hole, we brought back some geological trophies which had required hard work on our part. The larger of the two stalactites was turned over to the Oceanographic Laboratory of Miami, where Dr. Ginsburg was able to slice into it vertically and horizontally with a saw much more effective than our Rube Goldberg contraption. The most remarkable discovery he made was of thin layers of mud within the stalactite, containing marine fossils. These layers could have accumulated only once the stalactite was submerged, and they were therefore perfectly horizontal, like the level of the water. Now, however, they are at an angle of about 15° to the axis of the stalactite.

But not all of the stalactites were leaning. Some of them, as I have said,

Two specimens (Pomacanthus) of a species common in the coral waters of British Honduras.

were vertical. The latter had ceased growing when the level of the water rose about 12,000 years ago. Our expedition to the Blue Hole of Lighthouse Reef had provided proof that there had been an upheaval, followed by a period of stability.

Bob Dill is convinced that the slant of the reef corresponds to a shift in the earth's crust. And, in fact, British Honduras is located north of the Cay-

Calypso, preceded by one of our small craft, leaves the Blue Hole to make her way through the channel which leads out of the coral plateau.

man Ditch — which constitutes a major fault between the two Americas. This fault caused the drift which separates the Americas from Europe and Africa; and that upheaval was followed by a period of stability which lasted until the present — a period of 12,000 years.

Our little piece of rock is therefore an important witness, which allows us better to understand the drift of the continents in that area.

We offered the second stalactite to the government of British Honduras, which is supposed to make it a public monument in one of the public squares of Belize.

Another piece was sent to the Laboratoire de Faible Radioactivité at Gif-sur-Yvette, in France.

CHAPTER TEN

When the Water Boils

Flying over the Bahamas, even at a relatively low altitude, it is difficult to tell what is water and what is land or coral. The sky and the sea both are wrapped in a bright white vapor. Occasionally, one can make out points of coral protruding above the surface. A vein of deep blue water stands out for a moment, then is lost. There are dark, almost black spots — deep holes in the sea, none which appear to be very wide.

I was looking for the proper site for our next expedition. Here, as in British Honduras, there were Blue Holes — but in the Bahamas, they seemed smaller, more numerous, and, so far as I can tell from the air, much more difficult to reach than their Honduran counterpart.

The fact that there were many Blue Holes here immediately raised a question in my mind. I wondered whether they were interconnected by subterranean tunnels. This was followed by a conclusion of more immediate importance: the exploration of the Blue Holes of the Bahamas would be quite different from that which we had just completed at Lighthouse Reef.

It had been difficult enough for *Calypso* at Lighthouse Reef. I could imagine what it would be like here, where, so far as I could see, there was no channel, no way of access to the holes through an area which seemed completely filled with sand and coral.

As usual when confronted with a situation of this sort, my feelings were mixed. I disliked the thought of subjecting *Calypso* once more to the ordeal of picking her way through a maze of coral heads and shallows; and I was, as

(Above) One of the perils of navigation in the Bahamas: coral banks seem to be everywhere, at surface level, waiting for *Calypso*.

(Left) The Blue Holes of Andros Island, seen from the air.

always, conscious that one miscalculation, one moment's hesitation at the wrong time, could be fatal to *Calypso* — and perhaps to ourselves as well. At the same time, I had the utmost confidence both in *Calypso* and in the skill of her men. I suppose I might describe my attitude, when faced with a trial such as this one, as being one of optimism tempered by extreme caution.

The Bahamas comprise some seven hundred islands scattered across the sea and often connected to one another by "keys" — that is, by coral flats washed by the Gulf Stream.

The islands sit low in the water and are sometimes dotted with coconut palms or fringed with mangrove trees. They have changed considerably over long geological periods, but today most of them are bleak and barren.

The island of Bimini, to the west of Andros and almost directly opposite Florida, contains a "hole" which is rather large and which probably is very similar to the one at Lighthouse Reef. There is a vault and some debris on the bottom — but the ceiling of the cave is still intact and has not fallen. It is therefore not so much a "hole" as a cave, which is entered through an opening in the side.

Off the island of Andros, I flew over at least several dozen small blue holes. From the air, they looked like the footprints of a giant on the coral reef. They all looked alike, and they seemed to be laid out in a line of surprising regularity. Their small sapphire-blue circles stood out vividly in the emerald green of the surrounding reefs.

Four and one-half billion years of conflict among incredible forces on the surface of the planet — a planet forged in fire, chaos, and volcanic lava. And here was an opportunity to read that drama, not in the speculative volumes of the theorists, but on the very pages on which it had been written. Our feelings were not unlike those of a biblical scholar when he is presented with a fragment of papyrus from the first century and reads: "In the beginning God created the heavens and the earth . . ."

What we see today is the result of a very long process, in the course of which the earth's surface has been sculpted by the sea. The process is still continuing. The earth's appearance changes without our being aware of it. On a planetary scale, everything is moving: the coasts and the continents. The history of earth is a living history. Our seas are rising and falling, and everywhere there are scars which tell the story of earth's past. And that is what we went to the Bahamas to discover: indications of this everlasting process of change.

I knew beforehand that the sunken caves of the Bahamas would be more difficult to explore than the Blue Hole at Lighthouse Reef, and that it would require all the manpower and all the technical skill and equipment of *Calypso* if we were to achieve satisfactory results.

But I never doubted for a moment that our mission would be worth whatever effort it required. For it would be difficult to think of anything which epitomizes the history of the earth more cogently than those holes.

According to geologists, the Blue Holes of the Bahamas, which are similar to those found in the limestone country of Florida, were formed in the Pleistocene period. In Florida, there are stalactites and stalagmites 120 feet beneath the surface. If we could find stalactites in at least one hole in the Bahamas, we would have proof that these holes were once dry caves and that variations in the level of the sea were general and not due to local geological phenomena.

The holes originally were probably caves, or funnel-shaped holes. Rain water, which is slightly acid, destroyed the limestone which cemented the walls and caused cracks. Subterranean pockets were formed. At the end of the last ice age, the sea began rising once more and the caves were submerged, forming what are known today as the Blue Holes.

As I have already mentioned, these holes are also found in Yucatan, where they are called *cenotes*. The high priests of the Mayan civilization once erected their altars in these caves; and young children, who had been drugged beforehand, were sometimes thrown into the holes as offerings to the gods.

After a rather intensive air survey of the area, I concluded that the best place for our base of operations in the Bahamas would be the island of Andros. It is the largest of the archipelago, and it has the largest number of holes. These are in very shallow water, within the lagoon which lies between the reef and the coast of the island.

Fortunately for us, André Laban carried out a preliminary reconnaissance at Andros, as he had in British Honduras, and thus greatly facilitated the work of the whole team.

The voyage from British Honduras to the Bahamas had been a trek of 1,280 miles, much of it through coral reefs. *Calypso* and her men, as usual, had performed splendidly, expecting no favors from the sea and giving everything of their strength and their skill. In dealing with reefs, however, familiarity does not breed contempt, or even confidence. *Calypso* approached Andros with the utmost care. The island is about 115 miles long and is bordered on the east side by an astonishing fringing reef, which lies along the summit of one of the most spectacular escarpments in this area. This underwater canyon drops vertically to a depth of four thousand feet, and it is known as the Tongue of the Ocean.

The west side of the island is quite different. On our first visit, it seemed to us to be composed of a succession of shallow salt-water lakes separated from one another by mangrove-studded marshes.

The landscape is rather dreary, but it is more than compensated for by a seascape of incontestable beauty. Accustomed though we were to the coral splendor of the Red Sea and the Indian Ocean, we could not help admiring the extraordinary diversity of marine life in this area. The coral is quite near the surface, and we found it in combinations which were always different because the ingredients of the combinations were always present in different proportions. There were great violet and yellow sea fans, green algae, unusual staghorn coral, and massive brain coral.

The abundance of coral makes Andros difficult to reach and, over the centuries, has been the cause of many shipwrecks. Frequently, ships bound for Florida were blown onto these reefs. We found several of these ships. At Mayaguana Key, we saw one with its prow buried in the reef and its stern protruding into the open air. At the edge of the reef, the water is very deep; and on the wall of the reef we saw wrecked ships resting on ledges — ships of every kind and every age.

(Following page) Our diving team makes its way through an opening in the middle of the coral. The line which runs among the divers is a safety measure.

The subsoil of the Bahamas resembles nothing so much as a giant sponge. The entire archipelago is composed of the remains of a chain of limestone plains stretching in an arc about 80 miles long off the Florida coast.

The Blue Holes which I saw from the air lie in the middle of the reefs, and even on solid ground. But they are all surrounded by that long level reef which we had to cross with all our equipment.

There was something else which was going to complicate our mission: the fact that, at certain times, the water in the holes boils. Or at least it seems to boil. Falco, Yves Omer, and Jacques Delcoutère, when they were ready to go down for the first time into one of the holes, were taken aback to see the water begin to bubble and boil. Beneath the surface they could see no reason for this disturbance, but there seemed no point in taking any chances until it had stopped.

One of the peculiarities of the Blue Holes of Andros is that they are affected by the currents and countercurrents which are created by the tides. It was very impressive to watch the dark water in these holes foaming as it rose or fell from time to time with a great sucking noise. We could imagine ourselves being drawn down into unknown depths and finally smashed against the wall of an undersea cave. . . .

Bob Dill, who accompanied us to Andros, was fascinated by this phenomenon of alternating currents. The very complex system of the circulation of sea water through the Andros massif has never been fully explained; but the following theory seems to be substantially correct: The pressure exerted by the rising tide forces the water in the holes into the maze of tunnels, passages, and crevices which make up the subsoil of Andros Island. This water emerges inland, but near the shore, in accordance with the pressure of the rising tide. When the tide begins to go out, the water on the level reef then begins to drain off into the holes.

The rising tide and the current are not simultaneous; there is a delay between the former and the latter. When the tide begins to rise, the coral maze delays the flow of water; then, the water passes over the coral. When the tide goes out, the water on the reef tends to drain off through the holes, and this requires a certain amount of time. Therefore, the currents within the Blue Holes do not correspond exactly to the movement of the tides. When the tide begins to come in, water continues to drain out through the holes for another two or three hours.

Bob Dill noted that more water comes out of the ground than goes into it. I asked him how he explains this mystery, and he answered: "Because the system of tunnels under the island is much more complex than we think."

Bob also told me that the water which enters the holes from the sea is not clear. It is filled with plankton and animal waste. But when it leaves the holes it is crystal clear. Therefore, an abundance of organic matter has been deposited within the coral massif — which explains why certain old reefs are permeated with oil. At Andros, obviously, we were assisting at the formation of a future oil deposit.

In certain holes, fresh water rises and falls one or two inches, in a rhythm which more or less corresponds to the tides. This fresh water, because of its density, does not mix with salt water. Thus, when the tide goes out, the pressure at the far end of the subterranean system of tunnels diminishes. The fresh water in the holes forces the salt water through the cracks, fissures, and tunnels. Finally, the salt water emerges into the Blue Holes with such force that it seems to "boil."

Dark Corridors

Once we understood more or less how the Blue Holes were filled and emptied, we were in a position to begin working. We knew that the water was calm for a few minutes after the tides, and we concluded that we would have to take advantage of this relatively brief moment. Otherwise, we would run the risk of being dragged down into unknown depths from which we might never return.

In order to have as much information beforehand on what we were about to attempt, I asked one of our diving teams, led by Prezelin, to explore the network of tunnels which connects the Blue Holes. At slack tide, the divers went down — not without some apprehension — into the dark hole. They encountered a relatively weak tide, which nonetheless made their job much more difficult, for they were carrying some equipment: lamps to light their way in the depths, underwater cameras for taking photographs of anything of interest they might encounter, and finally the absolutely indispensable security lines with which to mark their route.

At the beginning of their dive, the divers were surrounded by coral of marvelous hues. They encountered a garden of sea fans inhabited by numerous fishes. Among these were snappers. Snappers are rather common fish, but here they were swimming on their backs. Instinctively, they turn their backs to the source of light; and in the holes the light comes from the bottom, where it is reflected from the sand.

In this topsy-turvy world, the snappers seemed perfectly at ease. More so than our divers, who are always aware of the danger of diving into caves, for, in the sea, a ceiling is a potential trap. If there is any malfunction of a diver's equipment in such a situation, then he is trapped, and his chances of survival are debatable — to put it optimistically.

The divers went down without too much difficulty. When they entered a narrow passage, however, they were surprised to discover that there was a current, although a weak one. In order to learn more about the ebb and flow of water within the Blue Holes, Raymond Coll installed an instrument in the passage which would measure the movement of the water over a twenty-four-hour period.

The divers found tunnels everywhere, crisscrossing one another, widening into chambers, narrowing into impassable holes. The darkness increased, and even our underwater lamps were too weak to bring out clearly the colors of the tunnel walls. The divers felt as though they were in a closed trap.

While our team was deep within the reef, the air bubbles from their breathing apparatus were passing through the porous limestone ceiling, through the thickness of the reef, and emerging in the water above the level reef. We were therefore able to follow the itinerary of our friends below.

Finally, our divers could go no farther. The passage had become too narrow, and their air bottles were scraping against the walls. They were forced to return, without having discovered any trace of the stalactites which we had hoped they might find.

In underwater caves which are lighted by reflection from the sandy bottom, fish swim bottom side up.

Despite the frustration of this first penetration into the black tunnels of Andros reef, I was determined that we would not give up. Day after day, we returned. And, as exhausting as these dives were, they were also exciting. At each fork in the tunnels, there was the risk of getting lost. At such moments, one tends to forget that time — that is, air — is limited, and that the least delay may have disastrous consequences. All of this, of course, adds to the adventure of exploration and to our sense of accomplishment when the day's work is done. But it also adds to the danger. And the most perilous aspect of it is that awareness of danger, like any emotion, wears thin after a certain amount of time, and a diver must struggle constantly to remind himself, from moment to moment, that he must never relax his vigilance.

Some passages were so narrow that the divers' air bottles were caught on projections from the walls — which in this case was a problem, since the passages were also too narrow for the diver to be able to turn around in order to disengage himself. In such instances, he had to walk backward until he reached a wider section of passageway.

Every day, our teams glided in and out of these tunnels and passages, but they found no stalactites. This is not to say there are none, for not all the tunnels — of which there are certainly over a hundred — were explored.

This fish seems perfectly comfortable despite its unusual position in the water. (Photo G. J. Benjamin)

A Collector of Blue Holes

We had with us at Andros a friend of André Laban's, Dr. George J. Benjamin, who is a chemist and a researcher. His true passion, however, is underwater exploration and photography. Whenever he is able to escape from his laboratory in Toronto, he rushes to Andros, where he has a boat and a small house. At that time, Dr. Benjamin had already discovered over a hundred Blue Holes. Of these, he had explored fifty-four — some of them to a depth of over three hundred feet. The unusual topography of these holes exerts a fascination on Dr. Benjamin which he finds irresistible.

Many people discuss the Blue Holes, but it seems that no one has had the curiosity, or the courage — or perhaps the means — to explore them. Only Dr. Benjamin has realized the three indispensable conditions for research of this kind.

To us, Dr. Benjamin was a companion as useful as he was amusing. He has a heavy Slavic accent, which gives an especially colorful flavor to his anecdotes. Moreover, he is a delightful and stimulating companion — a consideration of some importance when so many people live at such close quarters for a period of time.

For several weeks, he was our guide through the Blue Holes; but we never saw a hole with stalactites. Benjamin finally told us of a cave that he knew — the most magnificent cave, he said, and then went on to describe it in the most lyric terms.

He told us that a crab was responsible for the discovery of this cave, which he regards as the most important discovery of his career. "That day, I was with Archie Forfar, one of the best divers I have ever known. We had noticed a narrow opening located at the end of a whole series of Blue Holes. We entered the opening, and saw a very large crab. When Archie went toward it, however, it disappeared through an opening in the floor. In the light of our lamps, we could tell that the hole was very deep. We attached one of our emergency lights to a line and let it down. The line was two hundred feet long, but it never touched bottom. We could see the light shining far below us.

"We then secured the line and went down into the hole ourselves. At one hundred feet, we stopped. For all we knew, the hole was bottomless. But we decided to go on. Suddenly, we were in darkness. Our light, still far below us, had gone out. But we were in no danger. I had other lights with me. . . .

"I turned my camera toward the bottom and waited for Archie to point his electronic flash in the same direction. When the flash went off, for an instant — just for an instant — I thought I saw a great vault, from which

From left to right: André Laban, J.-Y. Cousteau, G. J. Benjamin discussing the next expedition. (Photo G. J. Benjamin)

massive columns were emerging — columns like stalagmites."

"Was it real," I asked, "or was it your imagination?"

"I could not be sure until I saw the photographs I had taken at that critical moment. And I was not disappointed. The photographs show we had discovered — quite by chance, and only because our light had gone out — the deepest Blue Hole into which man had ever dived in the Bahamas. They also showed that there were stalagmites there, confirming that these Blue Holes, like those of British Honduras, were formed by fresh water at a time when the level of the sea was much lower than it is now."

Shark Holes

Naturally, I was extremely interested in this account of Dr. Benjamin's, and I asked him to guide me to the cave which he had described with such obvious relish and such a wealth of detail. For almost two months we had been searching for such a cave, without success. And Dr. Benjamin's cave apparently had not only stalagmites, but enormous ones. He told me that

A diver makes his way through a difficult passage while pushing his lights ahead of him.

(Right) The whole exploration team is in the cave.

some of them must have been a hundred feet in length. "Unfortunately," he added, "we were not able to get close to them. All I could see with my light were greenish pillars. When we went down into that cave for the first time, I had no idea of its size; but, at first sight, it was most impressive. It was only later, after we had begun exploring it, that we realized what an exceptional place it was." *

Dr. Benjamin, whose accent increases in proportion to his enthusiasm for his subject, warned us that this cave from the Arabian Nights was guarded by sharks. These were only sand sharks, to be sure, which are inoffensive enough so long as they are not provoked. With their moustaches, they resemble large catfish; very large catfish, since they are usually between five and ten feet long. We have encountered sand sharks on several occasions, lying on a sandy bottom and blending into it (hence their name). But we had never disturbed them, and they had always ignored us. If we have learned anything about sharks in our years in the sea, it is that they are almost totally unpredictable. I say "almost" because we have discovered this: that if a diver leaves a shark alone, there is a chance that the shark will leave him alone. It had always worked for us in these caves — but we had no assurance that our luck would hold.

Dr. Benjamin's adventures with sand sharks had been more exciting. In order to be able to photograph the vast underwater chambers of his cave, he had synchronized his camera with three flash apparatuses carried by other divers. The divers carrying the flashes were swimming a few yards from one another, with electric lines running among them, when suddenly they saw two sharks in the beams of their lights.

"The sharks were surprised," Benjamin related, "and no doubt frightened also. They were heading straight for us, and then they made a sharp turn and plunged into the darkest corner of the chamber. They were so close to us that, when they veered away, we could feel the motion of water from their tails.

"Even in their corner, they must have felt that they were trapped between our lamps and the wall. The braver of the two rushed one of the divers, and then turned toward me. A moment later all that I could see was a cloud of brown mud enveloping our cameras and cables, the arms and legs of the other divers, and the silhouette of the shark. It was quite a spectacle. I

---

*Dr. Benjamin was quite disturbed by the idea that, after reading this book, some divers might be tempted to explore these caves. He pointed out that, during 1971 alone, more than twenty divers lost their lives while attempting to do so. It is therefore important for us to say here that such explorations are only for very experienced divers with the proper equipment, who are aware of and are prepared to cope with the dangers that await them.

pressed the lever of my camera, but nothing happened. The electric wires undoubtedly had been broken in the melée.

"We did not wait for the next attack, and we got back to the surface as quickly as we could. But when I developed my film, I discovered that there was a fairly good photograph which had apparently been taken close to one of our lamps."

(Following page) The bottom around the Bahamas is populated by coral, sea fans, and sponges, the equal in number and beauty of those in the waters of British Honduras.

CHAPTER ELEVEN

A Hole in the Wall

The east side of Andros Island is unusual in both its configuration and its vegetation. I had noticed this when I first flew over this area. One senses that it is not made for man. The sea is barely distinguishable from the land. And, in fact, there is very little land. Everything is filled and covered with living coral which multiplies unceasingly, extends its growths everywhere, and makes these islands almost inaccessible.

Living animals of all species are crowded in among the mangroves; and the mangroves, in turn, are perched precariously among the coasts and occasionally dip into the sea. Nonetheless, they grow and multiply.

This exotic world, seen from *Calypso*'s observation platform, presents a mixture of clear blue water, dark holes, and a slightly misty sky — all bathed in heat.

The dark, almost black holes in the landscape are regarded by the natives as objects of terror and, at the same time, of almost irresistible temptation. There are many fishes in them, and the natives are usually willing to use their boats in the holes; but they refuse under any condition to dive. They believe that the holes — which they consider to be "bottomless" — are inhabited by monsters; by evil spirits, known as *lusca*, which are always ready to drag men and their boats down into the bowels of the earth. These mon-

(Left) In the course of these difficult explorations, *Calypso*'s rear deck is encumbered not only by the two minisubs, but also by electrical cables, lighting equipment, coils of line and buoys.

Our Chaparral, an amphibious vehicle, made it possible for us to transport our equipment from *Calypso* to the Blue Holes.

sters resemble giant octopuses; and, if pressed, a native is usually found who can tell one the names of the latest victims to succumb to their voracious appetites for human flesh.

These people were filled with wonder and admiration at our courage in being willing to brave such dangers, and they never tired of enumerating the various fates which awaited us in the holes.

The real danger was from the boiling which occurs regularly in the Blue Holes and the strong currents which flow in with each tide. For six hours at a time, the sea spouts into these small holes like a spring; then, the water of the lagoon is sucked out. All of these hydraulic phenomena — which we believed extended to the very deepest underwater tunnels — formed whirlpools and a swell of which we were extremely wary.

It was clearly beyond our capability to explore the whole of the reef. It would have taken months merely to attempt a systematic coverage of all the Blue Holes. Dr. Benjamin himself, with his two sons, Peter and George, has explored only fifty-four of these underwater caves. It seemed best, therefore, for us to take Benjamin's word and to go down only into the caves which he regarded as the most interesting ones. Since we had already inspected those which lay close to the shore, we now planned to begin exploring those which lay farther out.

The most serious obstacle we faced was the ubiquitous coral. We had to cross the long level reef which belts the island; and, since it was impassable even with our Zodiacs, we resorted to more primitive means — we walked and we swam across it. No shoes were proof against the sharp edges of the coral; and the sea urchins and sea anemones, both of which are stinging animals, had no mercy on us.

We went from shallow pool to shallow pool in the reef, stumbling, staggering, somehow managing to stay upright and to retain our hold on the equipment we were carrying. When we reached the island's shore, we had to climb through the mangrove trees.

I could not help thinking that Andros Island must be very much as the earth itself was when the first bits of land began to rise above the sea and life began to develop upon them.

Prezelin finds a ray resting on the sand at the bottom of a pit.

The crossing of the reef did give me an over-all grasp of the underwater system under our feet within the reef, for it was obvious that the holes we saw corresponded more or less to the system underneath.

There was one way I could think of to trace all the ramifications of the network of tunnels in that gigantic sponge known as the Bahamas. I decided then and there that, later on, I would have our divers release a coloring agent — fluorescein — in one of the holes. We would then be able to see all the points from which colored water might flow on the surface of the reef.

One of our most difficult tasks in the Bahamas was to transport our equipment. At Lighthouse Reef, we had a good deal of worry and trouble maneuvering *Calypso* to our Blue Hole; but once we were there, we could handle our material — even the heavy pieces — with relative ease, for *Calypso* was moored directly over the Blue Hole. But in the Bahamas, everything would have to be carried ashore, and then carried from hole to hole. To make matters more difficult, there were hundreds of holes scattered all along the coral reef, and there were even some located rather far inland.

Fortunately, this was a problem that I had been able to foresee. Aboard *Calypso*, we had a vehicle designed for exploration both on land and in the sea. For, in addition to our scooters, our wet submarine and our minisubs, we had a small amphibious vehicle to carry ashore all the equipment we might need.

The Chaparral

One of our teams, led by Albert Falco, took the Chaparral, our amphibious vehicle, on its maiden cruise to Andros Island. Except that it was a cruise only in the widest sense of that term. The coral of the level reef was so dense and regular that, despite Falco's expert handling of the craft, Raymond Coll and Louis Prezelin had to jump into the water and guide the vehicle for several hundred yards, through a succession of pools and sharp coral formations. It was an ordeal which exacted an alarmingly large price from the tires of our Chaparral.

Things went no better once the vehicle had reached shore. The Chaparral, weighted down with our camera equipment, was bogged down on the shore; for the shore was nothing more than a marsh of stagnant water broken here and there by a few twigs.

The Bahamas, and especially Andros, cannot be described as attractive. Trees grow surprisingly well and pulpwood is a major export; but the natives, for a reason which remains a mystery, systematically burn all vegetation. We

Louis Prezelin and Raymond Coll, after their dive, will be picked up by Albert Falco in the Chaparral.

(Following page) At the bottom of the hole, our divers release a harmless dye so that we can discover the surface openings of this marine maze.

frequently saw the ashes of such bonfires; and here and there we ran across them still burning.

To the litany of our woes, I should add that the Bahamas has an enormous population of doctor flies so aggressive that, not being content to feast on us while we were ashore, they followed us out into the open sea. Their bites were quite painful; and they were joined in their attack by other insects so small as to be almost invisible — the "see-saws."

From the air, I had seen some perfectly round holes on the west coast of Andros, and I had decided that those were the holes to be explored. Unfortunately, very few of these holes were situated in such a way that they could be reached on foot; and, in any case, our equipment was so heavy that we could not possibly consider carrying it on our backs for any distance. Thus, I expected that the Chaparral would be very handy, despite our first somewhat disappointing experience with it, and I was rather proud of having thought of bringing it along.

But, even with the Chaparral, we had a very difficult time. Falco drove the vehicle, and he was compelled to pick his way with utmost care through a veritable obstacle course of natural impediments. Even then, he narrowly avoided sinking into a stretch of fetid mud. After countless detours, he finally came within sight of the Blue Hole which we had chosen for our dive. Prezelin and Coll, who were the Chaparral's advance party, cut their way through a

stretch of thorny bushes, while Falco did his best to follow them in his vehicle.

When our team reached the Blue Hole, it looked to them like nothing more than a small, calm pool. The surface was quite still, and there was no boiling. But Falco and his friends were aware of a phenomenon which Dr. Benjamin had often observed and of which he had been the victim at least once. It frequently happens that a layer of fresh water forms above the salt water. Rain water, being lighter than salt water, has a tendency to float on the surface of the sea, for example, and does not readily mix with it. A problem arises from the fact that a diver's buoyancy varies according to whether he is in fresh water or salt water. If he dives into one of these holes when it is covered by fresh water, he drops like a stone until he reaches the salt water — and then he bounces up as though he had landed on a trampolin. The separation between the two layers of water is sometimes as well defined as that between oil and water. This phenomenon occurs quite regularly after long, dry spells. Benjamin allowed himself to be taken unawares in a Blue Hole about 200 feet wide, south of Stafford Creek, almost in the center of Andros. The water was perfectly clear; but it was fresh water, going down to a depth of 125 feet, whereas the depth of the hole was only 200 feet.

Sulphur Water

The first thing that Falco's team did, therefore, was to make sure that the Blue Hole was indeed filled with salt water. The color and odor of the water were sufficiently characteristic for them to make no mistake about it. It was cloudy, and contaminated by decomposing organic matter. In such cases, the water generally contains no life forms. Nonetheless many of the Blue Holes are said to contain many fish, especially at certain times of the year. Fishermen come in great numbers to catch snappers and jacks.

The presence of hydrogen sulfide is explained by the decomposition of organic matter in the water. This gas is particularly abundant near the surface of the hole; but, as the divers go deeper, the water becomes clearer.

Most of the Blue Holes have the same general configuration. The upper part, which is sometimes perfectly round, looks like an immense barrel. The walls go down almost vertically; and usually, at a certain depth, there is an

(Right) A minisub has been equipped with floats to enable it to cross the coral barrier.

(Following page) The team goes into our "hole in the wall" on a reconnaissance mission. (Photo G. J. Benjamin)

overhang or balcony on one of the walls. Sometimes this overhang leads into a corridor, and the corridor may lead to a succession of horizontal chambers. In such cases, the opposite side is usually covered with mud and slants gradually toward this overhang.

Raymond Coll and Louis Prezelin found such an opening during their dive. They entered it and noted that there was a light current. At that depth, the water is clear; it must flow through the tunnels by means of which these underwater passageways are interconnected.

Coll and Prezelin began their exploration. But they had to proceed slowly, because of the weight of the equipment they were carrying.

Before them, they saw the corridor widen, then lead into a submarine chamber. A few fishes, frightened, scurried away into the darkness. Prezelin moved his light around until its beam came to rest on a large, motionless form lying on the sandy bottom. In the light, Prezelin could make out its shape: a ray, whose coloration blended in with that of the sand. The ray did not seem in the least frightened and made no attempt to swim away.

We concluded later that these chambers often serve as shelter — either hiding places or places of ambush — for large forms of marine life. Most often, these are predators waiting for the tide currents to bring them victims.

Demonstration by Color

Raymond Coll was of the opinion that the small cave which he and Prezelin were exploring was suitable for the experiment they were supposed to carry out. He therefore removed a small plastic bag from his belt. It contained a powerful coloring agent which, when released into the water, would presumably find its way back to the surface. This agent — which is green — had been specially chosen because it is completely innocuous to whatever flora and fauna may be found either under the ground or in the waters of the Bahamas.

The purpose of the experiment was to demonstrate that the Blue Holes are interconnected by tunnels — tunnels so small that our divers cannot enter them. I believed that the green dye would sooner or later emerge in the water over the level reef which we had crossed with such difficulty. I therefore had a launch cruise along the coast to observe the Blue Holes in the level reef and signal us when the green dye began to appear in the water.

The cave within which Raymond Coll released the coloring agent was about eight hundred feet away from the nearest Blue Hole being observed by the launch. We had to wait about twenty minutes — it seemed much longer —

before we were rewarded for our patience. Around the launch, the sea took on a light green color at first. A diver got into the water immediately, and he noted that it was quickly taking on a darker and darker hue as it rose from the bottom of the hole. The launch notified us by walkie-talkie of the good news. We now had proof that the Blue Holes are connected underground by a network of submerged passageways.

In order better to understand the topography of this complex network of tunnels, we repeated the experiment at several points on the reef. This gave us some idea of the extent of the network in the limestone beneath us and enabled us to compare these data with the results obtained from our measurement of the currents.

It goes without saying that Dr. Benjamin and Bob Dill were particularly enthusiastic over this new information on one of the most complicated geological formations, and one of the most mysterious, that exists simultaneously on both land and sea.

Plateaus in the Sea

The islands of the Bahamas are actually the summits of the largest range of limestone plateaus in the world. These are extremely high plateaus; for the distance from the tops of them — that is, from the islands — at slightly above water level down to the base of the range, far beneath the surface, is over five miles — 27,000 feet, to be exact.

Geologists tell us that the entire region is sinking at the rate of 33 centimeters (about 13.5 inches) every eight thousand years. This sinking is not noticeable, however, because it is compensated for by the sediment deposited on the islands and also by the coral, which grows at approximately the same rate.

An Unusual Excursion in the Minisub

One of the greatest of the many services rendered by Dr. Benjamin was to give us the benefit of his vast knowledge of the mysterious topography of the Andros reef. He had told us that near Stafford Creek, on the north side of the island, there was a hole which opened not toward land, but into the mighty wall which descends into the great depths of the Tongue of the Ocean.

This hole is located on a ledge about 125 feet beneath the surface, and, when the water was calm, even the simple act of diving was a pleasure. The

(Above) The minisub is positioned just above the "hole in the wall."

(Right) The minisub begins its descent into the hole. It has barely enough room to move.

(Below) A diver helps the minisub to squeeze into the hole.

Blue Hole into which we dived was some two hundred feet deep. At the bottom, it turns at a right angle and empties directly into the open water. About halfway down, however, there is an opening into the sea through which light pours. Twice in our dive, therefore, we were surprised to come upon patches of blue light emanating from the open sea.

There was much marine life along the walls of the Blue Hole — sponges of all colors; timid spirographs always on the verge of fleeing; Alcyonaria (pink or green soft coral). And finally, at the exit of the cave, the wall is covered with magnificent lavender sea fans.

A hammerhead shark hovered nearby, but it took little notice of us.

I wished to film this hole. We had seen nothing similar among the Blue Holes of Andros, or in the underwater caves which we had visited previously. Its greatest attraction, I think, lay in the blue light which flooded the narrow passage and which favored such a wealth of marine life.

Unfortunately, the extraordinary aspects of this hole — which Benjamin called his "hole in the wall" — began only at a depth of one hundred feet. This meant that we would have to carry all our material to that depth — our cables, lights, cameras, etc. — and then take it down even deeper, to a depth of 175 to 200 feet. We had already had our fill of hauling around heavy equipment. It occurred to us, however, that the easiest thing would be for us to send a minisub into the hole. It would be risky, of course, for the hole was rather narrow, the angle at the bottom quite sharp, and there was a chance that the minisub might be wedged in. Even so, I was determined to try. Using the minisub, we would be able to film with its two cameras, and its lights would provide ample illumination.

This was an exceptional use of a minisub. These vehicles were not designed for use in water accessible to divers, or for diving into pits, but for depths in excess of 325 feet. But the acrobatic aspect of my plan interested me. The minisub would have the opportunity to show us exactly how maneuverable it really was.

I was counting heavily on the help of the divers who would accompany the minisub and help it through the narrow stretches. I also planned to have a camera team hovering around the vehicle to film its exploit and show how difficult it was.

Albert Falco, our best pilot, was chosen for this experiment. He took the minisub down to one hundred feet, just above the narrow stretch through which he would have to pilot his vehicle. Around him, the divers and cameramen were trying to help, some of them directing Falco by means of gestures, others pushing or pulling the minisub so that it could enter at the proper angle. They were able to see that in the passage there would be very little space to spare between the minisub and the rocky walls of this almost perfectly cylindrical hole.

Slowly, the minisub dived. Occasionally, its sides scraped the rock or grazed a sea fan. Two divers, Prezelin and Yves Omer, and a cameraman, Claude Beconier, had gone ahead. They were the first to reach the blue light of the opening onto the open sea. Falco, who was in communication with us by telephone, let out a gasp of surprise when the minisub drifted down into that light.

Finally, the minisub had to undertake the most difficult part of its mission: the elbow, or right angle, at the bottom of the hole. We could all see Falco in our mind's eye, worriedly calculating whether or not his vehicle would be able to make such a sharp turn. But he continued piloting the minisub with his usual competence, turning the cameras and lights on and off according to the directions received from the gesturing cameramen and divers.

There was a tense moment as he reached the turn. But, with only a brief scraping of the wall, the minisub was through. Except for losing a few flakes of paint, there was absolutely no damage to it. Falco was now in the long corridor leading into the sea, and at the end of it he could see a flood of light. He inched forward carefully, for this final stretch was quite narrow, and its walls were encrusted with marine life. But soon, the minisub was in the open water. Its emergence was witnessed by a hammerhead shark, no doubt attracted by the bizarre animal we called our minisub.

The hole had been stripped of its mystery. Thanks to our film and our photographs, millions of people would be able to visit it without the slightest danger.

CHAPTER TWELVE

The Expedition of the *Scorpio*

Calypso's team had explored a large number of holes, caves, and caverns at Andros without discovering a single stalactite like those we had seen in the Blue Hole of Lighthouse Reef. It remained for Dr. Benjamin, who had devoted his life to the study of the geological mysteries of the Blue Holes, to lead us to what he called "my cave" — a large cave, difficult and even dangerous to reach, but which we hoped would offer us some reward for our efforts.

I entrusted the mission of organizing this, the most difficult exploration of my career, to my son Philippe.

It had all begun in a hotel in Miami. Dr. Benjamin had gone to visit Philippe. He was accompanied by a very engaging American diver, his son Peter (the discoverer of the main pit in 1967). The two of them had described at length the "fantastic caverns" where the water was of extraordinary clarity.

Calypso, at that time, was on an expedition. Philippe therefore rented the *Scorpio*, apparently the only boat available in Miami for a reasonable price. It belonged to a group of Cuban *émigrés* and, as Philippe described it, it was "sixty feet long, of which fifty were rotten." It had two engines, a generator, and water tanks; but nothing worked. When Philippe sailed from Miami he

(Upper left) An island covered with vegetation in the Bahamian archipelago (Joulter's Cay). Right: Tongue of the Ocean, with a depth of from 3,200 to 4,700 feet. (Photo Francoise Bourrouilh–Elf R. E.) (Lower left) The northeast coast of Andros Island. The geology of this island has been under study only for the past fifteen years. (Photo Francoise Bourrouilh–Elf R. E.)

was accompanied by Raymond Coll, Claude Beconier, a cameraman, Eugène Lagorio, François Dorado, and two Americans: Bob Dill, a geologist, and Tom Monnt, chief instructor at the University of Miami. Tom, as it happened, was also Dr. Benjamin's diving companion of long standing. Philippe realized that the compass had a 17° error. This was not very encouraging, particularly since the *Scorpio*'s mission was to take her wandering amid the coral reefs of the Bahamas. Somehow, following the Gulf Stream, the *Scorpio* reached Bimini. By that time, oil was running through the starboard engine as through a sieve. A pan had been placed under the engine to catch the oil as it leaked out; when the pan was full, the oil was poured back into the engine, and the comedy began again from the beginning. But that was the least of it. En route, the generator had caught fire and exploded. There had been ten or twelve short circuits in the electrical system. It was, in a word, a memorable voyage.

The team, as soon as it arrived at Andros, visited Dr. Benjamin and his son and daughter-in-law, whom they found in a small house near his prize cave. On land, they discovered that Dr. Benjamin's everyday garb was a pair of flowered pajamas and a hat bearing the legend: LONG LIVE FLORIDA.

Dr. Benjamin's experience and advice allowed us to gain a clear view of the problems we were facing and saved us a good deal of time. Every morning, he visited Philippe and his men aboard the *Scorpio*: and his approach was usually hair-raising. His boat would not go into reverse gear, and he had therefore adopted a unique maneuver for reaching the *Scorpio*. He would come at the ship at full speed, cut his engine when he was a short distance

(Far left) An American naval helicopter gives Captain Cousteau a lift to Andros Island.

(Left) Captain Cousteau brings the team some unexpected supplies.

(Following page) *Calypso*'s divers cannot resist photographing some of the particularly arresting landscapes they encounter.

away, then rush forward to try to cushion the shock with his arms. Often, he did not succeed. There were times when the men aboard the *Scorpio* thought their boat had been hit by a torpedo.

Like any good and prudent diver, Dr. Benjamin briefed our men thoroughly before taking them down to his cave. He particularly warned them against the mud which lay on the bottom and which would rise up in black clouds if it was disturbed. And, he told us, we must take special care to avoid brushing against any object we might encounter, no matter what it might seem to be.

The entrance to Benjamin's cave was not a vertical hole, but an incline; or rather, a sort of lip which began not far below the surface. It was perhaps sixty-five feet long and not more than six feet wide. Beginning at a 45° angle, it ended by being vertical — what mountain climbers call a "letter box."

Calypso's team had already dived there; but it had turned toward the north, where the passageway is stopped up. Benjamin took Philippe and his team toward the south.

In several spots there had been landslides, and the tunnel was very narrow. Usually, the men could find a way around the debris. It is likely that, even if the tunnel had been blocked, it would have been possible either to go around the blocked point, or perhaps to crawl under the debris; but in either case it would have been dangerous.

The passageway through which Benjamin led the team was eight hundred feet long, and led into an angled corridor. And nearby, in a semicircular chamber, they found the stalactites.

A Problem in Logistics

In order to film this labyrinth and the stalactite cave, our team needed light; at least two kilowatts of it. But all of the necessary equipment had been aboard *Calypso*, and all that Philippe had been able to do before leaving Miami was to purchase a large electric cable. He had assumed that he and his team would enter a vertical shaft to a depth of about 150 feet and that the corridor leading to the cave would be horizontal. In that case, there would have been no problem with equipment. But, in fact, the passage was slanted and contained many irregularities, and the corridor was winding and difficult to enter. This complicated topography made it almost impossible to handle heavy cables which are difficult to maneuver in the best of circumstances and which have a tendency to wedge themselves into the smallest crack or crevice. It would have been possible to lighten the cable by attaching it to floaters, but Philippe realized that it would then have merely risen to the ceiling. The fact was that the 1,000-foot cable was too heavy for a skeleton team of divers.

Finally, the team decided to use our underwater (pyrotechnical) torches for light while filming. Each torch, however, allowed only five minutes of filming; beyond that time, the smoke from the torches clouded the water. The bubbles from them rose to the ceiling, burst, and caused debris to fall, which further contributed to reduce visibility.

Philippe first spent a certain amount of time, in almost total darkness, deciding on the angles from which he wished to shoot, and he quickly realized that the circumstances for filming were not ideal. "The second time we started filming," he explained, "I was behind the other divers, who were moving toward the exit to find the nylon security line laid by Dr. Benjamin. By then, however, clouds of mud had risen in front of us and enveloped me. Since I was the last one in line, I was soon in total darkness. I switched on the battery-operated light on my camera, but all it did was to create a white area around me. I was completely disoriented. It was a very strange sensation. Moving along the walls, I was hunting desperately for the nylon cord and could not find it. In order to avoid being blinded by the glare from my camera lights in the milky water, I switched them off. Then, far in front of me, I saw a tiny light moving back and forth. It was Tom, the American diver, who, seeing that I was not with the others, had stayed behind and was signaling with his light.

"If I had not thought to extinguish my light, I would never have seen his signal; and since he, like the rest of us, was short of air, he would have been obliged to return to the surface."

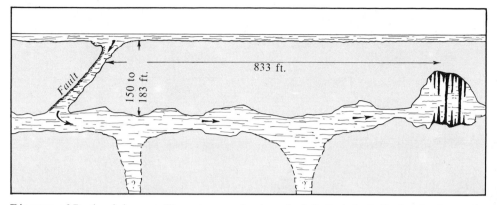

Diagram of Benjamin's cave. The entry opening is to the left, and the hall of stalactites to the right.

Filming in Quick Tempo

Philippe was determined that his film must convey the impression of a trap, a labyrinth in which one ran the risk of becoming lost at any moment. The risk, in fact, was proportionately greater because Philippe refused to tie himself to the other divers, for then he would not have been able to move ahead of the other divers or fall behind them with his camera.

During the three weeks of this expedition, the team dived twice every day. But each dive was of very short duration, because it could take place only during the brief period that the sea was either at low tide or high tide. In other words, there was a period of only twenty minutes allotted for each dive. The men stood around the hole, fully equipped, waiting for the water to stop boiling. Then they dived. Ten minutes were allowed for our team to reach the cave and film it, and another ten for regaining the surface. Usually, the sea began boiling again while the divers were at a decompression stop, lasting 40 minutes.

For a dive at 180 feet lasting twenty minutes, a diver must stop twice for decompression; once at twenty feet, and once at ten feet. This could only be done in the barrel-shaped hole leading to the bottom. When the water began to boil, the divers had to hold onto the security line, from which they fluttered like flags in the turbulent water. The divers, on one occasion, were caught in the current; but fortunately, it was the current of the incoming tide, and they were whisked up to the surface like so many corks. No one was injured. All they had to do was pull themselves down toward the base of the shaft and wait for the proper time to rise to the surface. If the current had been running

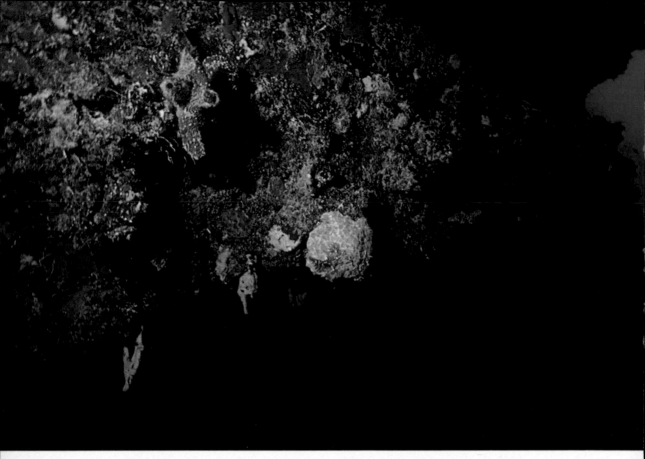

(Above) A wall of the cave, covered with animals and calcareous algae.

(Right) The passageway leading to the hall of stalactites. (Photo G. J. Benjamin)

in the opposite direction, or if the men had been in a horizontal corridor at the time, it is certain that it would have been impossible for them to reach the surface.

Every evening, the *Scorpio* docked at an Anglo-American base nearby (AUTEC, or Atlantic Undersea Test and Evaluation Center); for this complex of reefs is patrolled by the U. S. Navy. There are stations all along Andros Island, and at one of these, at South Bight, Philippe was able to take on fresh water. The seamen there were also kind enough to effect repairs on the *Scorpio* — particularly on the generator. We would never have been able properly to organize our dives if Philippe and his friends had been obliged to cross the Tongue of Ocean to find fresh water.

Risks to be Avoided

I was able to join Philippe's team at Andros for only three days; and only through the courtesy of the U. S. Navy, which put a helicopter at my disposal.

These three days gave us time for four dives: two at low tide, and two at high tide. I must say that these descents into the heart of the coral reef impressed me as being one of the most difficult things we had ever undertaken as a team.

Most of my impressions from diving into these caves are of danger, fear, and near-misses. I have always resolutely opposed anything which places our lives in danger. After all, we are not daredevils. In every undertaking, I insist that the risks be studied in detail; that everything be foreseen and organized in such a way as to reduce these risks to a minimum. In underwater caves like those of Andros, the slightest mishap can quickly turn into a tragedy.

One of the problems we had to face during the mission of the *Scorpio* was that of time; that is, the strictly limited time that we could remain in the caves. Philippe and Raymond Coll had calculated this carefully, allowing a margin for error.

Before my first dive to Benjamin's cave, the entire team stood around the hole, waiting for the water to stop boiling. I looked at the faces behind the divers' masks. Everyone was serious, silent. It was not a silence based upon fear, but rather one founded upon an intelligent appraisal of the danger involved.

We dived the instant the water was calm, down to the passage leading to the stalactite cave. Before that narrow crevice, I had a ridiculous thought: it was like being swallowed by a giant oyster.

From the moment a diver enters such a corridor, he must make optimum use of everything he has — his muscles, his heart, his mind. It is no longer possible for him to reach the surface in an emergency, as one is able to do in the open water. Surrounded by walls of rock, one is compelled to succeed.

I held on to the security line which Dr. Benjamin had permanently installed along the passage which goes down to 150 feet. I was tired from the trip to Andros, and, also, I was out of training. I had spent too much time in airplanes. My head ached, my ears hurt. But I could not slow down. There was already too little time. If I delayed the team and we were caught by the tide, all of us would be in serious danger. I glanced upward. I could see the light, dimly. The divers' bubbles, rising along the sloping wall, were silver in its glow. Then we were in total darkness.

We paused for a second as François Dorado shone his beam into the long tunnel before us. My friends around me were only dark shadows. . . . I struggled against the narcosis I felt enveloping me. The best remedy was action. As soon as everyone was in the tunnel, I gave the signal to move forward. I looked at my watch. We had only eight minutes to travel the eight hundred feet to the cave.

Above us, the ceiling was a constant reminder that we were not only underwater, but also underground. But below, we rarely saw the bottom, nor even rocks. There was only dark water and unknown depths — an abyss into which we would sink forever if we lost consciousness. I was keenly aware of how vulnerable we were in this alien element. In order to escape from the pervasive impression of unreality, I focused desperately on the thin cones of brightness dancing on the walls: the beams of our lights. Yet, my mind clouded over once more. We were in Dante's Inferno, and the gates were closing behind us.

Suddenly, the passage grew wider, and I knew that our exhausting voyage had reached its end. I paused to switch on my light. My heart was beating rapidly, and I was breathing too fast. I made an effort to breathe normally. If I exhausted my supply of air, I would never get back to the surface. In the dim glow, I could see that the other divers had also turned on their lights.

Before us was the cave. I could see stalactites — hundreds of them — hanging from the ceiling. They had been here, in darkness, for thousands of years. From the bottom of the cave, stalagmites rose to meet them. We were overwhelmed by the spectacle.

Acrobatic Cinematography

It may be that I was not sufficiently aware of the problem that I was creating for Philippe and his friends in their work. In my excitement, I went forward rapidly in the water. Philippe was filming the corridor behind us; then he had to catch up with me and pass me in order to get shots of me from the front. He went to the deepest part of the cave to film our arrival as we entered the cave one by one. These shots would have been impossible without some maneuvers on Philippe's part, which were true acrobatics. That is, they would have been impossible if Philippe did not have a very precise knowledge of what can and cannot be done in the water. He possesses an innate sense of movement in the water.

I glanced around and saw Raymond Coll, holding a loaded camera, floating gracefully in the dim light. I was struck by the smoothness of his movements. Like Philippe, he is so expert a diver that he achieves an unconscious and wholly natural elegance in the water.

The stalactites were before me, ready to reveal their history. None of these was slanting, as those of Lighthouse Reef had been. Apparently, the Bahamas were not subjected to an upheaval similar to that which affected the caves off British Honduras — not even before the melting of the great glaciers and the rise of the oceans which filled the caves 12,000 years ago.

(Above) The beauty and bright colors of the reef form a striking contrast to the darkness pervading the passageways of the cave.

(Right) Some of the stalactites, which we filmed for the first time, are of surprising delicacy. (Photo G. J. Benjamin)

All too soon it was time to retrace our steps. Once more, Philippe moved quickly ahead of us to film our return trip. Using two cameras, he shot a total of eight hundred feet of film during our brief journey into the bowels of the earth.

As we moved away from the cave, the liquid smoke from our torches clouded the water, and we no longer knew where we were, or what we were doing. I was in a trance, hypnotized, enchanted by what I had seen in the magic cavern.

Despite all our precautions, the sediment on the bottom had been disturbed by our comings and goings, and gradually visibility was reduced to zero. One by one, our lights were extinguished. Suddenly, the spell cast by the

sight of the cave was broken and was replaced by consternation. We must find our way out quickly; and we must be certain that no one is left behind. But we were surrounded by black, troubled water. After having felt our way back to the exit from the hall of stalactites, we began the long, exhausting swim back along the tunnel; but already we were in the grip of the ebb tide, and we were being pushed along at a more rapid pace than I had originally thought.

I felt that we had reached the base of the corridor leading out of the cave. The exit was above us. By touching my companions and pulling them toward me, I succeeded in grouping them; for my last torch had just gone out, and we were in total darkness. Nervously, I felt for the nylon security line — the lifeline which would lead us toward the light and to safety.

I touched it, grasped it. We started moving out of the cave, into the corridor; and then, with a sensation of enormous relief, we reached the hole leading to the surface.

But time was running out. We still had to make our decompression stops, and very soon the water would come boiling into the hole as the tide came in. As we waited patiently, we distracted ourselves by watching a familiar spectacle: several large jacks were having their parasites removed by wrasses. Each jack presented itself in the traditional position: leaning forward slightly, gills partially open. The wrasses were busy removing the tiny crustaceans which bothered their clients and which the wrasses themselves find so delicious.

Raymond Coll was carrying a rather surprising find: a sea fan from the very bottom of the cave. One would have thought that no fixed animal could possibly live in such darkness. The sea fan probably survived on plankton washed into the corridor by the tides.

We had now completed the second and final decompression stop, and we rose to the surface. The water was already beginning to boil. We saw the sun and the coral. A light breeze was moving across the reef. Our journey was over.

I shall never forget the marvelous sight of that cave. But it was with a sense of satisfaction that I returned to the world of men. I was exhausted, but content. We were now the possessors of a secret shared by only a few divers. We had lived through one of our most exciting and satisfying adventures. We had dived to the outer limits of reasonable risk, and we had survived because our team was expert, ready, and confident of its own abilities. And also because Philippe had foreseen all the risks and taken them into account.

We wanted to know still more about the underwater caves. And, in fact, in the entire adventure which is the unending voyage of *Calypso* through the

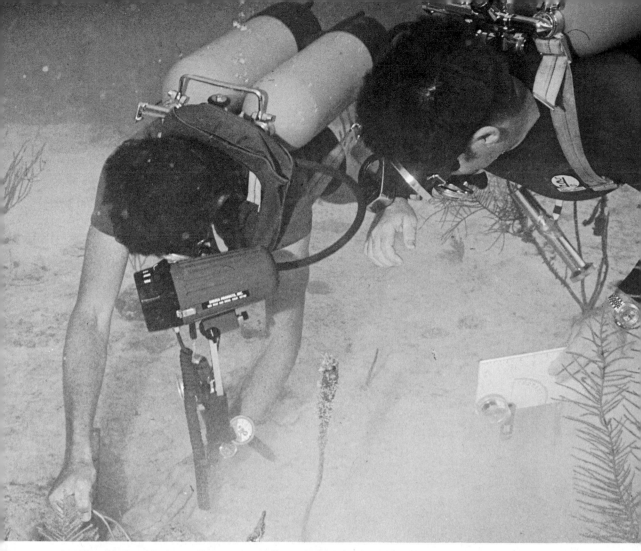

Bob Dill, with the aid of one of our team, makes a sketch of the cave.

(Following page) We were particularly impressed by the vaults of the cave, which are so deep that our strongest beams were not able to illuminate them entirely.

seas of the world, we have never been able to control our curiosity. Curiosity, then, is what motivates us and justifies us.

Christmas Trees

As Philippe said, "When we climbed out of that hole, we at least knew that we had done something."

He certainly had, especially in being able to film the dive so expertly in circumstances where there was no possibility of retakes. There had been no

time for such refinements; and, in any event, it had been my last dive into the cave, for I was leaving in a few hours. Philippe, however, was remaining, for he still had much to do. I wanted the film to show the smallest details of the cave, as well as the gestures of the divers, and even the courage required of the divers. Moreover, Bob Dill, who had explored the Blue Hole of Lighthouse Reef with us, had just arrived, and Philippe wanted to take him down into the cave to get his professional reaction as a geologist.

Philippe sometimes complained facetiously that Bob Dill was too quick with his hammer. He no sooner saw a small stalactite than he had it off the ceiling with one stroke of his hammer and into the little bag which he always carried with him.

The American divers generally were a surprise to us. First of all, we were astonished and impressed by the amount of equipment they carry. Bob Dill, especially, was always covered with instruments of various kinds when he dived. In addition, when he went into the water, he always towed a load of supplementary equipment. There was a tablet for taking notes; an ax; string; an emergency regulator; and a life jacket, which is used to give the diver additional buoyancy — a necessary convenience in view of the load of equipment which he carries. What struck us most about the American divers, however, was not their gear, but their great physical courage and their determination.

I noticed that American divers wear a kind of mask which cuts down on visibility. These masks have a groinlike arrangement around the nose, which is apparently intended to dispose of water which enters the mask. The result, however, is that American divers usually have a reduced field of vision in the water.

It is true that the *Scorpio* team played a few unpleasant practical jokes on some of their very capable American guests. One of the chief victims was Dusty Way. On one occasion, Dusty was standing on the level reef, with water around his legs, lowering some lights at the end of a cable. The lights were operating off the boat's 220-volt electric system, but the system was not grounded. Suddenly, Dusty shot straight out of the water. He had been bending over at the moment of the shock, and he bounced up like a spring and landed ten feet away from where he had been standing. Happily, he was wearing a diving suit, which protected him against the coral.

(Right) Aboard the *Scorpio*, Captain Cousteau with his son, Philippe, prepares to dive.

Our divers felt a great sense of relief when they finally emerged into the open water and the rays of the sun . . .

(Right) . . . but we will never forget our fantastic underwater cathedral. (Photo G. J. Benjamin)

The whimsical operation of our underwater torches also caused several tragic-comic incidents. It often happened that they did not light right away. In such cases, the divers did not throw them away, but kept them in their hands or under their arms. Almost invariably, the torch would light two or three minutes later and send a spurt of flame into the diver's mask.

The more Philippe and his team familiarized themselves with the labyrinth at Andros, the more they came to appreciate its beauty and its infinite complexity. They began to meet some of its inhabitants also. Among these were a few sharks, who quickly vanished. No doubt, they knew of passages into the stalactite cave other than the one we had used. There were also several lobsters. Philippe noted that one part of the wall of the cave had a reddish look to it; but there was no biologist with him at the time, and there was no way of identifying the fauna encrusted on the wall.

Philippe decided to concentrate on Benjamin's cave. He had come to know it fairly well, and it was the most spectacular of those we had seen. The others seemed all very much alike, although Bob Dill dived into them several times and came up with some new data on the complex geology of the Bahamas.

During the three weeks of the *Scorpio*'s mission at Andros, Philippe and his team made more than thirty dives. Finally, it was time to leave that desert region to which no tourists flocked and no boats came. Our men took their leave of the American sailors who had so generously placed the resources of their AUTEC base at their disposal. Then the *Scorpio*, its hull disintegrating, its engines sputtering, and its compass still awry, sailed for Miami.

It was the end of one of our most dangerous and rewarding adventures.

Calypso, meanwhile, had already set out in search of new adventures.

(Following page) The headlights of the minisub stir up marine life in the darkness.

ACKNOWLEDGMENTS

We owe a special debt of gratitude to Dr. G. J. Benjamin, who not only guided us in the submerged corridors of Andros, but also gave us the benefit of his remarkable observations and allowed us to reproduce several of his excellent photographs.

We are equally indebted to Mme. Françoise Bourrilh of the Laboratory of Historic Geology of the University of Paris and *agrégée* of the University, who has graciously shared with us the results of her special study of the geological formation of the Bahamas.

The photographs reproduced in this book were taken by Henri Alliet, Dominique Arrieu, Jéan-Jerôme Carcopino, Ron Church, Bernard Delemotte, Isabelle Deloire, François Dorado, Marcel Ichac, André Laban, and Jacques Renoir.

Some of the photographs taken on the surface were chosen from the personal collections of members of *Calypso*'s team.

Iconography: Marie-Noëlle Favier.

APPENDICES

Appendix I

CHARLES DARWIN AND THE GALÁPAGOS

Charles Darwin was born in England, in the town of Shrewsbury, in 1809. His father, Robert Darwin, was a physician; and his grandfather, Erasmus Darwin, had also been a physician and, in addition, a poet of some repute in his time. When Charles was ready to enter the university, the strong liberal, Protestant atmosphere in which he had been raised had exerted its influence, and he had decided to study for the Church.

Thereupon, as he described it, he "wasted three years at Cambridge." Although the course of studies which he pursued was intended to prepare him for the ministry, he showed a far greater interest in the natural sciences than in theology, and he spent much of his time collecting and studying insects, butterflies, and birds.

In his final year at Cambridge, Darwin gave up entirely his ecclesiastical ambitions and enrolled at the University of Edinburgh to study medicine. Once there, however, he discovered that he could not endure the sight of blood; and he spent his time at the university principally in the study of geology, under the celebrated Adam Sedgwick. He returned once more to Cambridge to take up his theological studies; and, in 1831, he was graduated with honors.

In the same year, upon the recommendation of John S. Henslow, Darwin's professor of botany (and also a clergyman), the young man was offered the position of naturalist aboard the H.M.S. *Beagle,* a small ship which was about to set sail on an extended mission of research and exploration. Although the post of naturalist carried no salary, Darwin was delighted at the prospect. Despite his great interest in the natural sciences, he had given no serious thought to making a career of them; indeed, he had abandoned all hope of becoming proficient in any branch of science, and he had resigned himself to the life of a clergyman of the Church of England.

The *Beagle* was under the command of one Captain Robert FitzRoy, a zealous Protestant, who was delighted at the prospect of having aboard a naturalist who was also a future clergyman. He was confident that a man of science who was also a man of God, or at least of the Established Church, would be able to interpret any discoveries made by the *Beagle* in conformity with the Bible. One of the duties of the *Beagle's* naturalist, in fact, was to work within the context of Genesis and to search for traces of the Deluge. Darwin's own approach to his work, at this time, was not at variance with that of Captain FitzRoy. He had already been exposed to the evolutionary theories of Lamarck, but, as yet, he saw no necessity for questioning the commonly accepted interpretation of the biblical texts concerning the origins and development of life.

The *Beagle* sailed from Plymouth on December 27, 1831; and from the moment that she moved slowly out of the port until her return five years later, Darwin was overcome with seasickness. It did not help matters that the *Beagle* was a ship of only 242 tons, and that they were so cramped for space that he had to share a cabin with Captain FitzRoy and with FitzRoy's collection of twenty-two clocks arranged in rows on shelves. For five years, Darwin's days were spent among the ticks and gongs of FitzRoy's prizes; and his nights were spent with his feet in a chest of drawers, for the cabin was so small that he could lie down at full length only by removing the drawer at a level with the foot of his cot.

The ship crossed the Atlantic and called at San Salvador, then dropped anchor at the port of Bahia where Darwin got his first astonished glimpse of a tropical forest. Then, moving down the eastern coastline of South America, the *Beagle* entered the bay of Rio de Janeiro.

At Punta Alta, Darwin discovered the fossilized bones of giant animals: the *Megalonyx,* the *Scelidotherium,* the *Mylodon.* What struck him most, however, was not the size of these creatures, but the fact that they were similar in some respects to existing species. The discovery of the skeleton of a horse — which had disappeared in the Americas and was unknown there when the conquistadors arrived — particularly astonished him, and it was the occasion of his first discussion with FitzRoy on the literal authenticity of the Bible and the reality of the Deluge.

At Tierra del Fuego, where the *Beagle* remained for a time, Darwin occupied himself with studying the natives and compiling a dictionary of their language.

After rounding Cape Horn in very rough weather, the *Beagle* arrived at

Valparaiso on July 22, 1834. Darwin took advantage of the location to scale the Andes and, at an altitude of over 12,000 feet, he discovered a number of marine mollusk fossils. This was proof that, at one time, the Andes themselves had been submerged and then, at a comparatively recent geological epoch, had been raised above sea level by some upheaval of the earth's crust.

The Galápagos

On September 7, 1835, the *Beagle* put out of the port of Callao, in Peru, and set its course for a group of volcanic islands called the Galápagos. They had been discovered in 1553 by Tomás de Berlanga, bishop of Panamá, and, at the time of the *Beagle*'s voyage, they were much frequented by whalers who came in quest of the giant tortoises which they used as a supply of fresh meat aboard their ships. (The islands had taken their name from these animals, which were found in great abundance there. *Galápago* is the Spanish word for tortoise.)

"Here," Darwin wrote, "both in space and time, we seem to be brought somewhat near to that great fact — that mystery of mysteries — the first appearance of new beings on this earth."

The *Beagle* anchored first off Chatham Island (San Cristóbal), at the eastern end of the archipelago, where the crew had its first glimpse of marine iguanas. The ship cruised among the islands for a month and, during that period, Darwin had the opportunity to dissect an iguana and analyze the content of its stomach. He was thus able to determine that these reptiles fed only on algae. He also captured three young tortoises which would eventually reach England alive.

Most of his work, however, was devoted to birds, and he was struck by their lack of fear of man. On James Island, he found twenty-six species; and he soon discovered that the species he had observed on that island did not exist on any other island of the archipelago or, for that matter, anywhere else in the world. They were similar, it is true, to other species found in the Americas; but only similar, not identical.

He also observed that birds of the same species differed from island to island. The finch (*Geospiza strenna*, which was later to become known as "Darwin's finch") was a perfect example of this phenomenon. The color of this species varied from black to green according to whether a specimen lived on an island of lava rock or one on which the vegetation was abundant. In the former case, the finch's beak was highly developed and hard, to enable it to feed on seeds and nuts; in the latter, the beak was long and narrow, to enable the finch to feed on insects.

Darwin was now on the track of a discovery which was to upset the then accepted ideas on the origin and stability of species. The Book of Genesis itself, so dear to Captain FitzRoy's heart, was to be cast into doubt; and, as can be imagined, the discussions between Darwin and the captain became lively indeed after Darwin had begun to draw tentative conclusions from his observations.

From the Galápagos, the *Beagle* set sail across the Pacific and called at Tahiti, New Zealand, and Australia. In the spring of 1836, she began the crossing of the Indian Ocean, and her first port of call was in the Keeling Islands. It was there that Darwin began to formulate the solution to a problem which had long occupied his mind: the formation of the atolls and the growth of coral.

On April 29, the expedition reached Maurice Island. Two months later, it rounded the Cape of Good Hope, and then called at St. Helena and Ascension Island. The course had already been set for England when Captain FitzRoy decided that, before the *Beagle*'s mission could be said to be complete, one more trip into American waters was necessary. The expedition therefore set sail for Brazil, and called at Bahia and Pernambuco.

It was not until October 2, 1836, that the ship finally dropped anchor in the English port of Falmouth, after a voyage of five years.

Darwin was now twenty-seven years old. He would live to be seventy-three, but he would never again set foot outside Britain.

Immediately after landing, Darwin set to work to classify the numerous specimens collected in the course of the *Beagle*'s voyage. The events of the expedition itself were set down in Darwin's *Journal of the Voyage of the Beagle*, which was published in 1839 and created a sensation.

Darwin married a cousin named Emma Wedgwood. Unlike many marriages of the era, it was a love match and Darwin settled down happily with his wife in Kent, sixteen miles from London. It was there that he wrote the five volumes of *The Zoology of the Beagle*.

In 1837, in his first group of notes on the evolution of species, he had written: "Since last March, I have been studying the American fossils and the species peculiar to the archipelago of the Galápagos. The facts concerning the species of that archipelago are at the source of all my ideas."

In 1838, he had read Malthus' *An Essay on the Principle of Population*, first published in 1798, and was inspired to begin assembling the elements for a work on natural selection. He was perfectly aware that his ideas would provoke a scandal in Victorian England, where the literal truth of the biblical account of creation was accepted by society at all levels, and it was to be twenty years before he ventured to make his theories public.

Darwin's theory of the evolution of species provoked many caricatures in the British press. Here are two of them, both of which appeared in *Punch* during 1861. (Copyright Mansell Gallery)

THE LION OF THE SEASON.

In 1858, the opportunity presented itself for such an undertaking. In June of that year, Darwin received from Alfred Russell Wallace a copy of the latter's essay entitled, *On the Tendencies of Varieties to Depart Indefinitely from the Original Type*. The concepts put forward by Wallace in this work were similar to those held by Darwin himself since his visit to the Galápagos in 1835. Darwin warmly recommended Wallace's paper to his scientific colleagues, and it was decided that the two naturalists would read a joint paper before the Linnean Society of London.

One year later, Charles Darwin decided to publish the work which was to be acknowledged as his greatest: *On the Origin of Species by Means of Natural Selection, or the Preservation of Favored Races in the Struggle for Life*. The 1,250 copies of the first printing were sold on the day of publication.

Darwin's expectations of a great storm of scandal were fully realized. When the third edition of his work appeared, a general meeting of scholars and clergymen was convoked at Oxford for the purpose of debating the question of the origin of species. The clerical forces were headed by a man widely respected in Britain: Samuel Wilberforce, Bishop of Oxford. Wilberforce had made no secret of his intention of demolishing Darwin's theories; but those theories were defended on this occasion by men at least as capable as Wilberforce himself: Professor Henslow, Darwin's former teacher, along with T. H. Huxley and a botanist named Hooker. The climax of the debate came when Bishop Wilberforce asked Huxley whether he was descended from a monkey through his grandfather or his grandmother.

To his neighbor, Huxley said: "The Lord has delivered him into my hands!" And, to Wilberforce, he answered: "I should certainly prefer to be descended from a monkey than to be an educated man who places his talents and his eloquence at the service of prejudice and error."

The good bishop was shocked into silence. There was a great uproar in the hall. A woman fainted. And the erstwhile captain of the *Beagle*, Vice-Admiral FitzRoy, rose and brandished the Bible, shouting: "Here is the truth, and nowhere else."

After the *Origin of Species*, Darwin wrote eight books, the best known of which is *The Descent of Man*. His reputation grew with each new publication despite the acrimony of the debate which raged about his theories.

Neither fame nor notoriety caused Darwin to slow his pace. "When I am obliged to give up my observation and experiment," he had written, "I shall die."

He died on April 17, 1882, having worked until two days before, and was buried in Westminster Abbey.

Appendix II

THE PRE-COLUMBIAN CIVILIZATIONS OF THE ANDES

From the end of the last ice age, the Andes were inhabited by groups of humans who lived in caves. Traces of the existence of such groups have been found at altitudes of 12,000 feet. In the caves at Chilca, for example, fourteen successive archaeological layers have been found indicating the presence of humans over a period of some ten thousand years.

It is surprising that man should have chosen to live, and to remain, at such an altitude and in a region which is especially arid. It is possible that he did not live under those conditions for the entire year. The discovery not only of the bones of land animals but also of blue clams from the Pacific has led some observers to conclude that these groups lived along the shore for several months of the year and migrated into the mountains during the summer.

There is strong evidence of land cultivation in the lower Andes from about 6,000 years before Christ, and it is certain that cotton and kidney beans were known to the people living there at that time. By 1500 B.C., corn was being cultivated, and the potato had been known for many centuries.

The Kidney-Bean Planters
It seems that between 1800 and 1500 B.C., there occurred a considerable social and technical progress among these peoples — a phenomenon which F. A. Engel, in his *The Pre-Columbian World of the Andes*, described collectively as the formation of "societies of kidney-bean planters." Engel found villages

composed of small huts and also of more impressive buildings, as well as traces of ceramic pottery. This civilization disappeared suddenly and mysteriously between 500 and 400 B.C.

It was followed by a people who migrated to the Andes, perhaps from the Caribbean, and built circular houses the lower halves of which were sunk into the ground. They also constructed underground rooms in circular form, on the perimeters of which ascending rows of benches were scooped out of the soil. These were probably used for assemblies or for ceremonial purposes.

The staple of this culture was corn. The llama was considered sacred and was not eaten. Its wool was sheared and woven into clothing, and the animal was used as a beast of burden. For meat, the people of this civilization depended upon the guinea pig.

In the vicinity of Lake Titicaca, as well as at Cuzco, archaeologists have found pottery over three thousand years old. It has not been possible, however, to ascertain precisely what peoples were living in that area at that time.

Some authorities believe that the Urus, who still live on the floating islands of Lake Titicaca, are descendants of the Arawaks who emigrated from the Antilles — a group of scattered settlements which still exist in the Amazon Valley. It is believed that the Arawaks came from the north and crossed the virgin forest, bringing with them images of the serpent and the jaguar.

According to another authority, Dr. Jehan Vellard, an expert in Uru culture, the inhabitants of the floating islands are a mixture of Uru and Aymara. The only pure-blooded Urus who remain are those who live solitary lives in the mountains.

The Chavin Civilization

The first great civilization of Peru which can be clearly dated is that known as the Chavin civilization. It has been established by the radiocarbon method that, in the middle of the eighth century B.C., this new culture appeared in Peru already in possession of the technological and artistic skills which characterized it and which produced vast architectural complexes, vases in the form of human heads (these were probably portraits), and massive gold ornaments.

The experts are in disagreement as to whether this culture developed in Peru itself or elsewhere. Some believe that it had its origin in groups which migrated from China during the Chou dynasty, while others think that it was of Japanese origin. What is certain is that Chavin art has many similarities with the Mexican art of Oaxaca and Vera Cruz. It is also possible to see a relationship between Chavin architecture and that of the Mayas in Yucatan and Petén.

The Chavin civilization vanished suddenly toward the end of the fifth century before Christ, as mysteriously as it had come.

The Andean Societies

The Andean civilization which followed the disappearance of the Chavin culture was characterized by the existence of numerous regional societies, all of which made extensive use of copper, and all of which produced ceramics of different kinds and often of great beauty.

One of the most famous of these cultures is the Mochica, famed for its vase portraits and for a pink pottery ornamented with painted scenes.

Several excavations have shed light on various peoples who attained a relatively high degree of civilization between 500 and 200 B.C. — the Gallinazo, the Maranga, the Lapa Lapa, and the Paracas and the Nazca. (Fabrics and pottery of unusual quality have been found in the burial places of the two latter cultures.) These groups, for the most part, lived on or near the Pacific coast of South America.

Tiahuanaco

In the immediate environs of Lake Titicaca, at an altitude of over 12,000 feet, there is an immense architectural complex which at one time served as the source of inspiration for an artistic, stylistic, and technical culture which covered an enormous area. "Nothing is known," writes Professor Engel, "of the origins of the Tiahuanaco civilization. Radiocarbon dating of the lowest layers of the site indicate that they go back 2,000 years; and even these layers have yielded fragments decorated in a manner which reveals a high degree of proficiency."

It is likely that the ensemble of buildings found some 12 miles from Lake Titicaca was only one of the centers of a civilization which covered much of the Andes. Other sites have been found, the structures of which bear an easily discernible resemblance to those of Tiahuanaco.

The area ruled by the people of Tiahuanaco was not as large as present-day Peru; yet, the cultural impact of Tiahuanaco — in the form of artistic style and subjects — is evident in the Lower Andes and in the coastal regions, beginning at the middle of the eighth century A.D. Along Lake Titicaca, on the other hand, beginning at approximately 1000 A.D., it seems that little of cultural or artistic value was produced.

The site of Tiahuanaco, which survived the Incas and even the Spanish invaders, has unfortunately been ruined by successive demolitions. It is possible that some of the material was used for the construction of Incan monuments; most of it, however, was hauled away to be used in the construction of churches in the surrounding villages.

Of what remains, one can easily make out three large temples and a building the lower half of which is underground. These are constructed of enormous blocks of stone, cut, polished, and placed with rare perfection. They are, in Engel's words, "megalithic monuments which had no equal in grandeur in pre-Columbian America."

The most typical construction in this complex is a heavy monolith known as the Gateway of the Sun. It is ten feet high and four feet wide and is decorated in a style evocative of that which had come down from the heights of the Andes. The placement of this portal, isolated as it is from any other structure, constitutes one of the many riddles of Tiahuanaco. A number of large monolithic statues have been discovered in the vicinity of the Gateway of the Sun, and these are similar to the portal in style. Most of them are now in La Paz.

The Island of the Moon and the Island of the Sun in Lake Titicaca are both very ancient religious shrines. One of the conquistadors, Pedro Sancho, records that in the Temple of the Sun "there were 600 Indians in constant attendance, and a thousand women occupied in preparing the *chicha* which was poured in sacrifice over the sacred rock called Titicaca." (Quoted by A. Métraux, in *Les Incas*.)

Even today, 60,000 Indians go in pilgrimage every year to the sacred shrine of Copacabana.

The Aymara tribesmen, who are found today along the shores of Lake Titicaca, are the descendants of a people who were enslaved by the Incas and later rebelled against them. Some archaeologists believe that the Aymaras were responsible for the spread of Tiahuanaco civilization by means of an era of expansion which preceded the coming of the Incas. This hypothesis has been contested by other experts.

The Incas

After the collapse of the Tiahuanaco civilization, there is a hiatus in the history of the Upper Andes. Little is known of the events between A.D. 1000 and 1400, when the first information on the Incas becomes available.

It is known that there was a Chimu kingdom which succeeded that of the Mochicas between A.D. 1100 and 1200, and which existed contemporaneously with the semilegendary reign of the first Incas. The people of Chimu were experts in agricultural irrigation; and their capital city, Chan-

chan, was undoubtedly the largest city in the New World before Columbus' arrival.

The chief god of the Inca kings was the sun, whose image appears on the Gateway of Tiahuanaco. But they also prayed to Viracocha, the "creator-god" who had risen from the waters of Lake Titicaca to teach men the technique and the art of government. Some authorities believe that Viracocha was the supreme god of the mysterious nation to which we are indebted for the monuments of Tiahuanaco and of which the Incas were the heirs.

Appendix III

VARIATIONS IN THE LEVEL OF THE OCEANS

The relative levels of sea and land were long regarded as fixed and constant. It was not until the twentieth century that this static concept of the elements which make up our planet was abandoned.

Changes in the level of the oceans are brought about by several factors. The most obvious of these is the constant erosion of the land, by means of which elements are deposited in the sea and cause the water to rise. This, however, is responsible for only a small difference in water level.

A more important factor — and one pointed out by Darwin — are the glaciers which successively (and especially during the Quaternary Era) covered the surface of this planet. Carbon-14 tests have demonstrated that changes in climate, and the presence of glaciers in the Northern and Southern hemispheres, coincided with the lowering of the level of the oceans, while periods of relative warmth corresponded to a rise in the water level.

During the last glacial era — known as Wurm in Europe and as Wisconsin in America — the whole of Scandinavia, northern Germany, northern Russia, and northern Asia were covered by a sheet of ice. In North America, almost the whole of Canada was covered. One can imagine the immense volume of water which must have been taken from the oceans in order to constitute such a body of ice.

It has been possible, by a series of more than 80 tests with radioactive carbon, to establish a curve indicative of the level of the ocean along the coast of the United States over a period of some 35,000 years. Fifteen of these tests were concerned with changes which occurred more than 15,000 years ago.

The elements tested were shallow-water mollusks, oolites, coral algae, beachrock, and marsh peat. The results obtained indicate that the level of the oceans 30,000 to 35,000 years ago was approximately what it is now. Then, 16,000 years ago, during the last ice age, that level fell at least 400 feet. The water began to rise again during the Holocene Era, approximately 14,000 years ago, and continued to do so at a fairly rapid rate for some 7,000 years. The results of most of the tests taken in other parts of the globe are in agreement with the results of the curve thus established, indicating that this curve truly reflects the changes in the level of the oceans over the entire planet during this same period.

Several other factors may have affected the level of the seas. Temperature has a double effect: cold causes the molecules of sea water to contract, and the level may have been lowered in this way by several meters. Also, the accumulation of ice at the poles and on the surrounding land removed a great quantity of water from the seas — approximately five million cubic miles of it. This would have lowered the level of the oceans by some 650 feet with respect to its present level. In these conditions, the weight of the ice may have caused the continents themselves to sink.

Changes in the level of the oceans are therefore the result, not of a single factor, but of several complex factors in combination. The results of these changes are equally complex and are seen in major modifications in the configuration of both the continents and the oceans. Straits were bridged, islands were combined or united to continents, and so forth. All in all, the lowering of the water level resulted in an 8 per cent increase in the size of the continents themselves, while the continental plateaus were considerably reduced. The salinity of the oceans was increased slightly, by 35/1000 or 36/1000.

Appendix IV

CALYPSO AND HER CRUISES

Calypso, our oceanographic research ship, is a former World War II minesweeper. She was built in the United States in 1942 for the British Navy, and, after the war, I was able to buy it at Malta, thanks to the generosity of a British patron, Mr. Löel Guinness.

As ships go, *Calypso* is not very large, measuring 140 feet in length with a beam of 24 feet. She is, nonetheless, a solidly built vessel of 329 tons, with a double hull of wood, double planking, and very narrowly spaced timbers. She is marvelously easy to handle, and her shallow draft enables her to maneuver in and out of treacherous coral reefs with a minimum of trouble. Her two engines and propellers give her a speed of ten to ten and one-half knots. With a little squeezing here and there, she can accommodate thirty people.

The squeezing is not because *Calypso* is too small, but rather because she carries so much equipment, all of it necessary for marine exploration.

It was necessary, of course, to make extensive alterations before the minesweeper could become an oceanographic laboratory. We added a false stem, among other things, and a well which goes down about eight feet below the waterline and in which there are five portholes. From there, we can see— and film—what is going on in the water, even when *Calypso* is in motion. In addition, a double mast of light metal was built as far forward as possible on the deck. It serves as a radar antenna, as a sort of upper bridge from which to observe and direct a difficult passage, and as a crow's-nest from which we can observe the larger marine animals.

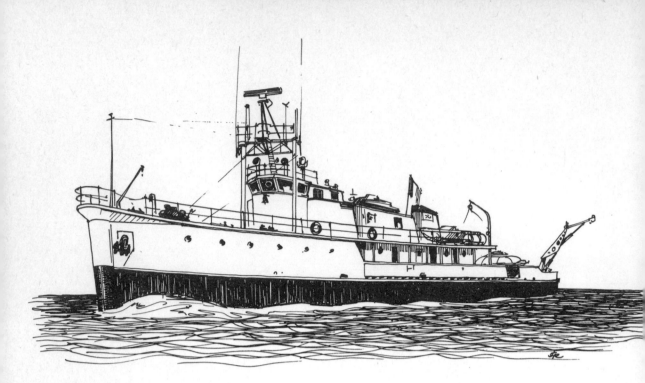

"CALYPSO"

Calypso, a former minesweeper built in the United States, has been modified several times to make her over into an oceanographic-research vessel.

Calypso's equipment is unequalled by that of any other oceanographic vessel: about twenty scuba outfits; a brace of underwater "scooters"; a "wet submarine"; two miniature submarines which we call "diving saucers" or "minisubs"; numerous small craft, such as our non-sinkable metal runabouts, our inflatable rafts (the Zodiacs), and our life boats which inflate automatically when placed into the water (the Bombards)—and an excellent collection of outboard motors. In order to be able to film properly both on the surface and in the water, as well as from our underwater craft, we carry fifteen underwater cameras, six commercial dry-land cameras, and a variety of special automatic cameras for filming in the deep. Then, since underwater filming requires an enormous amount of light, we must have a huge number of waterproof lighting fixtures and miles of special wiring. We have a closed-circuit-television system, by means of which we can see everything that happens either on the ship or in the water. We have an ultrasonic telephone for communicating with the minisubs and the divers, and for the divers to be able to communicate with each other. We have tape recorders and underwater microphones for recording the whole chorus of noises that pervade the "silent world" of the sea. And, of course, we have all the latest navigational aids: automatic pilot, radar, navigational sonar, and also a special sonar for use in

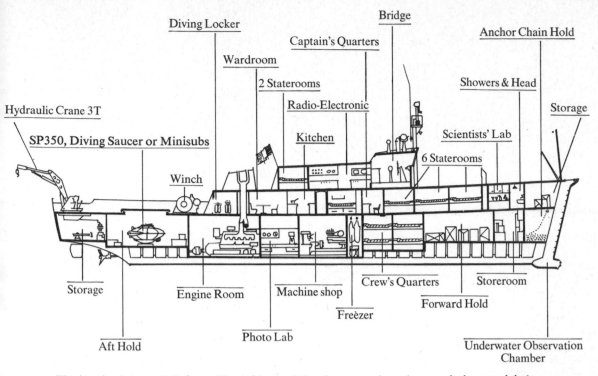

Diving Locker

Bridge

Anchor Chain Hold

Captain's Quarters

Wardroom

2 Staterooms

Showers & Head

Radio-Electronic

Storage

Hydraulic Crane 3T

Kitchen

Scientists' Lab

SP350, Diving Saucer or Minisubs

6 Staterooms

Winch

Storage

Engine Room

Machine shop

Crew's Quarters

Storeroom

Photo Lab

Freèzer

Forward Hold

Aft Hold

Underwater Observation
Chamber

The interior layout of *Calypso*. The bridge and the chart room have been entirely remodeled. Cabins were added forward. The rear hold is now used as a garage for the minisub.

very deep water. We also have several laboratories and aquariums aboard— one of the latter being an anti-roll aquarium, which remains stable (and therefore assures the comfort of its inhabitants) even in the roughest weather. And I won't even mention the cranes, hoists, compressors, etc., which are the ordinary equipment of any ocean-going vessel. What is surprising, therefore, is not that *Calypso* is crowded, but that she is able to carry all that she does, and still have room left over for thirty people.

Calypso's expeditions are not financed by any public or private subsidy. The Océanographic Museum of Monaco, to which we refer frequently in this series of books, lends *Calypso* its support; but it is scientific, not financial, support. *Calypso* is sponsored by a foundation set up in 1950: *Les Campagnes Océanographiques Françaises*. The sole funds available to this foundation derive from book and television royalties, royalties from industry, and fees for research undertaken on behalf of various commercial enterprises.

Calypso's first expedition, her maiden voyage, was made in 1951, to the

(Following page) The itinerary of *Calypso*'s longest cruise, which began in February 1967, and ended in September 1970. During this period, *Calypso* sailed some 140,000 nautical miles without once returning to her home port, and we shot 24 marine films for television.

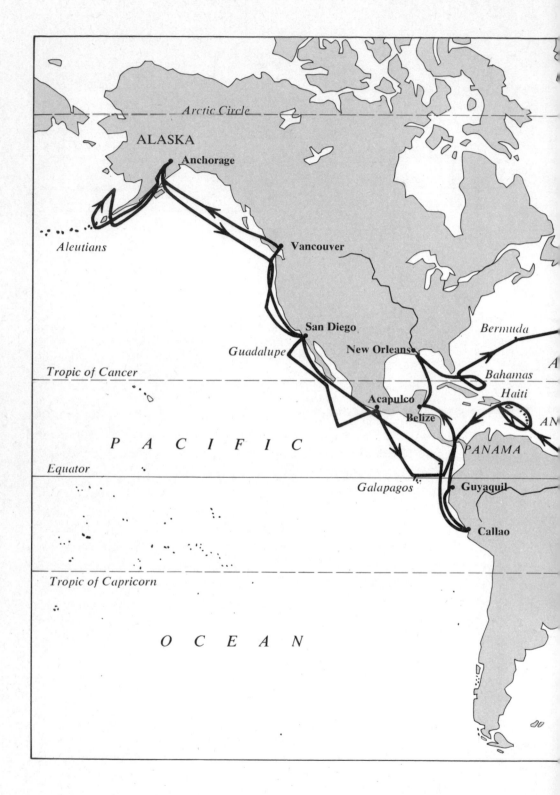

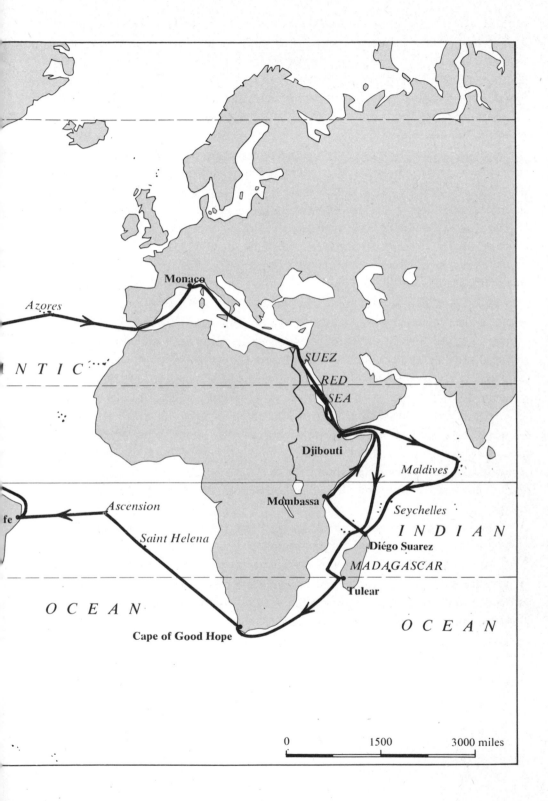

Azores

NTIC

Monaco

SUEZ
RED
SEA

Djibouti

Mombassa

Maldives

Seychelles

INDIAN

Ascension

Saint Helena

Diégo Suarez

MADAGASCAR

Tulear

fe

OCEAN

Cape of Good Hope

OCEAN

0   1500   3000 miles

Red Sea. Her next mission was in the Mediterranean off Marseilles, at Grand Congloue, where she participated in an archaeological dig, the subject of which was a Roman ship of the third century B.C. Since then, she has travelled through the Indian Ocean, and through the Atlantic to the Antilles, shooting films—notably *The Silent World of Jacques Cousteau*—and on scientific missions. Among the latter was an exciting photographic expedition, undertaken in collaboration with Professor Harold Edgerton of the Massachusetts Institute of Technology, to study various depressions in the ocean floor.

Calypso's expeditions are not reckoned in weeks or months, but in years. Her last major expedition lasted from 1967 to 1971 and took her 140,000 nautical miles across the Mediterranean, the Red Sea, the Indian Ocean, the Atlantic and the Pacific, northward to the Bering Strait. In the course of that voyage, we shot two dozen films for television.

After a brief layover at Marseilles (during this "layover" we shot several more films in the Mediterranean), *Calypso* set out on a new adventure: a lengthy expedition into the Antarctic, calling, en route, at ports in Argentina, Patagonia, the Falkland Islands, and Tierra del Fuego. The purpose of this expedition, which set out on September 29, 1972, is to study the effects of commercial hunting and chemical intoxicants on the warm-blooded animals most vulnerable to such abuses: blue whales, humpback whales, right whales, sei whales, finback whales, killer whales, seals, penguins, and albatrosses. For this particular undertaking, *Calypso* received a load of new equipment. Among the most notable items are a Hugues 300C helicopter, which can be disassembled and stored below deck; a helicopter landing pad on her stem; a hot-air balloon; special equipment for diving in Antarctic waters; new cameras; new lighting equipment, and so forth. And yet, *Calypso* does not complain.

The reader may have noticed that, in these books, we do not speak of *the Calypso*, but simply of *Calypso* as though she were a person. The reason is that she is something more than a ship to us. She has carried us to places where no man has ever set foot before. She has taken us safely through storms and hurricanes and coral reefs that would have sent a lesser ship to the bottom. She is our home, the center of our lives. And, in a very real sense, we depend upon her for our lives. By now, it is difficult for us to think of her as a *thing*. What must we call her? The ancient Greeks, I think, had the answer to that question, as they did to so many others. According to Homer, the original Calypso was a nymph of the sea who held men in a bondage which abounded with delights. That description seems to fit our own *Calypso* very well.

ILLUSTRATED GLOSSARY

ILLUSTRATED GLOSSARY

ALCYONIANS

Alcyonium is a genus of Cnidaria, class Anthozoa, subclass Octocorallia. The true alcyonians are what divers call "soft coral." They are colonial and tend to form huge clusters of individual polyps. The polyps have eight tentacles and eight internal divisions, and they are united by means of canals.

Alcyonians have no skeletons. Instead, they have stiff, calcareous spicules.

Soft coral is remarkable for the beauty of its colors — pinks, greens, and blues — and for the fact that at night the coral swells to huge proportions.

ALEUTIAN ISLANDS

A group of islands off the southeastern coast of Alaska and separating the Pacific Ocean and the Bering Sea. The Aleutians were discovered in 1741 by Vitus Bering, a Danish explorer.

ALLIGATOR

The alligator is a vertebrate reptile of the family Alligatoridae. There are two principal genuses: the *Caiman*, which is found in tropical and equatorial America; and the *Alligator*, found in the southern United States and in China, along the lower Yang Tseu Kiang.

Alligators live principally along rivers and in swamps, and feed both on vertebrates and on crustaceans and insects.

The American species, *Alligator mississipiensis*, has been known to attain a length of eighteen feet. It has been widely hunted for its hide, and exists now in greatly reduced numbers.

ALPACA

A mammal of the Camelidae family, the alpaca is found in the cold, mountainous regions of South America. It is a hybrid obtained by the Peruvian Indians by crossing the vicuna with the guanaco. The alpaca's wool is highly prized because of its length and quality.

AQUA-LUNG

The Aqua-Lung, or self-contained underwater breathing apparatus (SCUBA), was designed in 1943 by Jacques-Yves Cousteau and an engineer, Émile Gagnan.

The principal characteristic of this apparatus is that it is an "open-circuit" device; that is, the used air is expelled directly into the water, and fresh air is provided not in continuous fashion, but whenever the diver inhales.

The air itself is stored in one or more air tanks (or "bottles" or "cylinders") which are strapped onto the diver's back. Its flow is controlled by a regulator, which delivers air when the diver inhales and which assures that the pressure of the air corresponds to that of the water surrounding the diver. When the diver exhales, the used air is fed into the water by means of an exhaust located under the hood of the regulator. Two flexible tubes run from a mouthpiece to the regulator; one is for inhalation, the other for exhalation.

This simple and safe apparatus, entirely automatic and easily mastered, has, in effect, opened the doors of the sea to man and made it possible for a large segment of the public to experience the thrill of diving. The invention of the Aqua-Lung therefore was a decisive step forward in man's conquest of the sea, and even in the history of human progress.

ASCIDIANS

Ascidians — more generally known as sea squirts or tunicates — are chordates. The sea squirt is, in effect, a small water bag, yellow or red or violet, which is free floating as a larva but becomes fixed at the adult stage. There are two openings: one to take in sea water, from which it removes oxygen and the tiny organisms on which it feeds, and the other to "squirt" the waste water.

Despite their primitive appearance, ascidians have gills, a stomach, an intestine, and a V-shaped heart. The contractions of this heart move the blood first in one direction, then in the other — 80 pulsations in one direction, 40 in the other. Ascidians are hermaphroditic.

BARRACUDA

The barracuda is a well-known flesh eater of the tropical seas. In appearance, it somewhat resembles the pike, with its prominent teeth, well-defined jaw, and its elongated body the color of polished steel.

The largest of the barracudas is the Great Barracuda, which sometimes reaches a length of over six feet. It prefers to travel in groups of three or four when fully grown. Smaller barracuda are often found in schools comprising individuals all of the same size and of the same age or generation.

Barracudas have a bad reputation. In certain areas, they are even more feared than sharks. This reputation is probably due, at least in part, to their ferocious appearance, razor-sharp teeth, mean-looking eyes, and general behavior. On the whole, however, it has been my experience that the barracuda's behavior is more dramatic than dangerous.

BERING STRAIT

A stretch of water connecting the Arctic Ocean and the Bering Sea. It separates the northeastern tip of Asia and the northwestern extremity of North America.

During the winter, the Strait — which is approximately 55 miles wide and 200 feet deep — is completely blocked by ice.

BLACK CORAL

Black coral is a hexacorallium, class Anthozoa, subclass Geriantipatharia, order Antipatharia. It is a colonial animal and its colonies are similar to those of sea fans, although the individual polyps are different. It has a hard skeleton.

BOOBY

The booby is a large marine bird belonging to the order Steganopodes. It is very common in the Galápagos and lives in colonies of several hundred individuals. It fishes by diving into the water and attacking a fish from underneath. Its prey is usually swallowed beneath the surface.

The female hatches her eggs by covering them with her feet.

There are three species of booby in the Galápagos: the blue-footed booby (*Sula nebouxi*), the masked gannet (*Sula dactylatra*), and the red-footed booby (*Sula sula*).

CENOTE

A natural depression in a limestone surface, deepened by the flow of underground waters. There are many such in Yucatan, in Mexico. The Mayan city of Chichén Itzá (meaning "the mouth of the wells of the Itzas") took its name from the three large cenotes nearby. One of these, with a diameter of almost two hundred feet, was a sacred well into which human sacrifices were thrown. The victims were most often young girls, covered with flowers and jewels and rendered unconscious by drugs.

CORMORANT

The wingless cormorant (*Nannopterum harrisi*) is found only in the Galápagos Islands. It is the only cormorant incapable of flying. Yet, despite its name, it is not wholly wingless. It has retained stubs of its wings, which it uses to maintain its balance in the water.

The wingless cormorant is an excellent swimmer and finds its food in the rich waters of the archipelago. It drinks sea water and has a special gland which allows it — like the iguanas — to filter out the excess salt.

CURRENTOMETER

A device perfected by the Center for Advanced Marine Studies of Marseilles which records the strength and direction of currents.

DARWIN'S FINCH

These finches (Geospizinae) were the species which inspired Darwin to formulate his theory of natural selection.

The finches arrived in the Galapagos from the American mainland at a time in the distant past. They disappeared completely, but left in their place some twenty-six different species which have survived on the different islands of the archipelago. It is likely that the islands originally had no land birds.

The search for food has resulted in the evolutionary modification of the beaks and behavior patterns of the finches. Their other characteristics have remained unchanged so far as most of these species are concerned.

The oldest of the present-day Geospizinae is a grain-eating finch which lives on the ground and whose beak is not unlike that of the sparrow. From that rather primitive type there have developed birds with much stronger beaks, capable of breaking open very hard nuts.

Two other species have developed elongated beaks, but these are equally strong and enable the birds to feed on the flowers and fruits of cactus plants.

Other species with moderately long beaks eat fruit, buds, and insects.

There is a particularly differentiated species, *Camarhynchus pallidus*, which digs into the trunks of trees to find insects. It uses the spine of a cactus plant, held in its beak, to bore into the trunk, and is the only bird known to make use of a tool.

DOLICHOCEPHALIC

A dolichocephalic creature is one whose cranium or skull is long in proportion to its breadth, as opposed to a brachycephalic creature which has a rounded cranium.

A dugong.

DUGONG

A marine mammal of the order Sirenia. Its body, with horizontally flattened tail, resembles that of a cetacean.

The average length of the dugong is about ten feet, and it weighs between 400 and 500 pounds. It is found in the Indo-Pacific region, from the Red Sea to Australia.

The dugong lives alone, or in small groups, and feeds on algae and water plants which grow in shallow water.

ELECTROCARDIOGRAM

The report delivered by an electrocardiograph which indicates, by means of electrical impulses, the vital functions of the heart.

FALSE NOSE

Calypso's "false nose" is one of the modifications effected when she was transformed from a mine sweeper into a marine-research vessel. The false nose is a metal well attached to *Calypso*'s stem and descending to a depth of eight feet below the water line. There are five portholes, through which we are able to observe and to film what is going on beneath the surface even when *Calypso* is in motion.

FLUORESCEIN

A coloring agent, with a phthalein base, giving a very brilliant yellowish-green fluorescence to its alkaline solutions.

FRIGATE BIRD

A marine bird belonging to the order Pelecaniformes. It is always found in the Galápagos in the company of boobies. Two species are found there: *Fregata minor* and *Fregata magnificens*.

The wingspread of the frigate bird is over six feet. It is a powerful flyer, and it has a long, hooked beak. Its feet are quite small and are without webbing.

The frigate bird is a vicious fighter and will attack any intruder into its territory. It often seizes the prey of other birds. It does not dive, but is expert at catching flying fish "on the wing."

The male frigate birds are black, with brilliant green feathers and a red pouch under their throats. In the mating season, the male inflates this pouch. This process takes about twenty minutes; then the male waits quietly until a female spots him from the air.

GUANACO

A land mammal of the Camelidae family, the guanaco is the only member of the family still found in its wild state in the plateaus of the Andes.

The guanaco lives in the mountains of central Peru and in Chile, and is also found on the plains of Patagonia. It has practically disappeared in Argentina.

It is reddish-brown in color, with a pale underside.

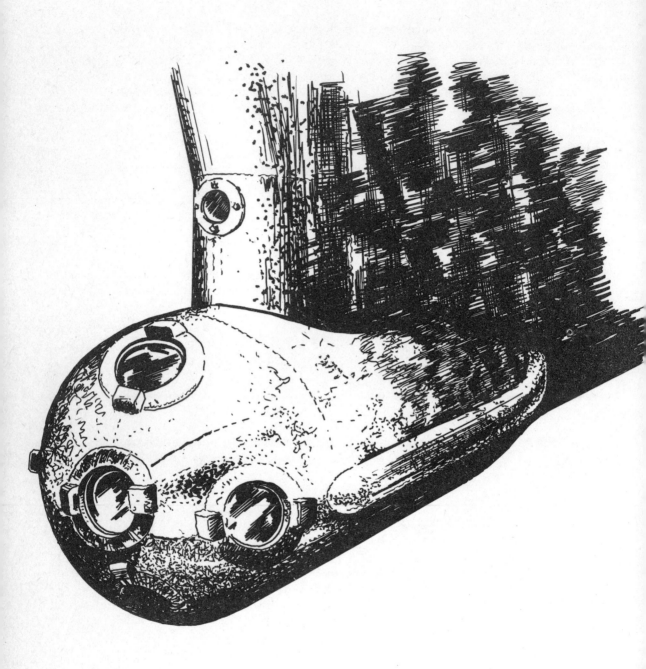

Calypso's observation chamber has five portholes. It is reached by means of a metal staircase attached to the length of *Calypso*'s stem. We call this whole complex "*Calypso*'s false nose."

GYROPILOT

A gyroscopic compass which automatically activates the rudders.

HAMMERHEAD SHARK

A fish of the class Elasmobranchidae, family Sphyrnidae, with a flat head which is elongated into protrusions on both sides. The eyes are located on the outer tips of these protrusions.

The *Sphyrna zyguenu* is found in almost all warm seas, and it is often seen in coastal waters. It is a powerful swimmer and feeds mostly on sting rays. It bears its young live.

A hammerhead shark.

HAWSER

A line made up of several strands which is used in towing and securing a ship.

THE HUMBOLDT CURRENT

The Humboldt Current, or the Peruvian Current, is a flow of cold water from the Antarctic running along the length of the South American coast. It joins the equatorial current.

KARSTIC

Relating to the karst, or limestone plateau, affected by chemical erosion.

KELP

Kelp is the common name given, particularly in the United States, to various large algae, most of them pheophytes, which are found off the coast of California and Mexico. They are also found near New Zealand, Argentina, Chile, and Peru.

Macrocystis pyrifera reaches a length of almost 200 feet. Its feet are firmly anchored in the bottom, and its strands float by virtue of air-filled bladders. It grows very rapidly. Some authorities believe that this species grows to a length of 1,000 or 1,500 feet.

Along with several other algae, *Macrocystis pyrifera*, *Pelagophycus porra*, and *Eisenia arborea* are known generically as kelp.

KEY

A coral bank. The word first appeared in its French form, *caye*, in Furetière's dictionary in 1690. It is taken from the Spanish *cayo*, meaning reef or rock.

KODIAK

An island off the state of Alaska, situated in the Pacific.

LEERFISH

Lichia amia, a member of the Carangidae family, is a beautiful fish which reaches a length of six feet and moves about in schools. It is a curious animal, but a prudent one, and apparently likes the company of divers. It is found both in coastal waters and near reefs, as well as in the open sea. Its hydrodynamically perfect form and its tapering body, ending in a crescent-shaped tail, give it a characteristically elegant appearance.

LEVEL REEF

The level reef is a coral reef found in shallow water in the tropics. It is a more or less long and unbroken plateau running along the coast, or along the summit of a submerged reef in the open sea.

LLAMA

A mammal of the Camelidae family found in the Andes from the equator to Tierra del Fuego. It is used by the Indians as a beast of burden and as a source of meat and wool.

MANTA RAY

Also known as the horned ray, the manta or devilfish has two protrusions from its head which resemble horns. It is one of the giants of the sea, and the larger specimens have a wingspread of about twenty feet and weigh as much as 1.5 tons.

The manta ray, despite its formidable appearance, feeds only on plankton. It absorbs its food by swimming with its mouth held open as it moves majestically through the water.

The manta ray is a relatively harmless animal, but the gigantic leaps out of the water of which it is capable, its size, and its strength make a fearsome adversary of it.

Two kinds of manta ray have been observed in the Indo-Pacific region, but it is likely that both may be included in the same species: *Manta birostris*.

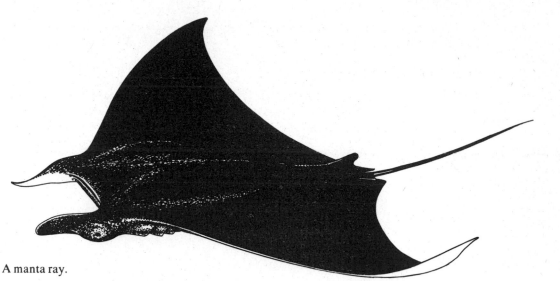

A manta ray.

MINISUB, or DIVING SAUCER

There are several types of minisubs, or diving saucers, designed by Captain Cousteau and developed by the Center of Advanced Marine Studies of Marseilles.

-The *SP-350,* a two-passenger vehicle, is equipped with a cinematographic camera, a still camera, a hydraulically operated pincer and lift, and a storage basket. It has been used in more than 600 dives.

-The *SP-1000,* or sea-flea, carries only one man but is designed to be used in conjunction with a second SP-1000. It has two exterior cameras (16mm. and 35mm.), both controlled from within, and tape recorders for recording underwater sounds. It has been used in over 100 dives.

-The *SP-4000,* or Deepstar, is capable of diving to 4,000 feet. It was built for Westinghouse and was launched in 1966. Since then, it has participated in over 500 dives. It is a two-passenger vehicle, with a speed of three knots.

-The *SP-3000* was built for CNEXO. It attains a speed of three knots, and carries three passengers.

MOCKINGBIRD

The mockingbird of the Galápagos Island is only slightly different from the mockingbird of the United States — but different enough, nonetheless, for it to have a genus of its own: *Nesomimus.*

It is about the size of a thrush, and often has black markings on the sides of its head, which is gray-brown on top and cream-colored underneath. It has a long tail and a curved beak.

The mockingbird is a predator, feeding on young finches, small lizards, and insects. It also eats the eggs of other birds. It is a very lively bird, and it glides, with its wings outstretched, rather than flies.

It lives in groups, and, as soon as one of the group finds something to eat, the others congregate and take it away from him. The strongest takes the prize.

Despite its small size, the mockingbird of the Galápagos is afraid of nothing; not even of man.

OBSERVATION TOWER

Calypso's observation tower or platform is a double-masted structure of metal placed as far forward as possible on her deck. It is, in fact, a sort of upper deck which serves as our radar station and is also very useful as a vantage point for observing the sea around us, for spotting various marine life forms, and for picking our way through dangerous reefs.

OPUNTIA

A genus of cactaceous plants having flat joints studded with tubercles bearing sharp spines. The tubercles bear flowers, mostly yellow.

PADIRAC PIT

A chasm located in France, in the Lot *département*, which consists of a large circular hole which empties into channels dug into the limestone by the flow of rainwater.

PENGUIN

Penguins belong to the order Impennes, and to the Spheniscidae family. They are perfectly adapted to marine life. Their feather-covered wings are used for swimming.

The penguin is able to survive in extremely low temperatures and to do without food for long periods. It feeds on fish, crustaceans, worms, and mollusks, and is found in antarctic and subantarctic regions. Its presence in the Galápagos is explained by the presence of the cold waters of the Humboldt Current.

PLANKTON

A complex of small organisms which live in suspension in sea water. There is the vegetal plankton, called phytoplankton, and the animal plankton, called zooplankton.

POMPANO and JACK

Pompano and jack belong to the family Carangidae, which is related to the Scombridae (tuna and mackerel). They normally live along the coasts, in tropical and temperate waters, but they are also often found far out in the open sea.

This family is a uniformly handsome group, usually having blue or pale green backs and gold or silver sides, which resemble polished metal. There is a well-defined lateral line running to the deeply forked tail.

PRIVATEER

A privateer was an adventurer, hardly distinguishable from an ordinary pirate, who preyed on Spanish shipping in the Antilles during the sixteenth, seventeenth, and eighteenth centuries, and occasionally upon the Spanish colonies in the New World.

Opposite: Our "observation platform" is actually a double mast installed on *Calypso*'s forward deck. We use it to observe marine life and to pick our way through reefs.

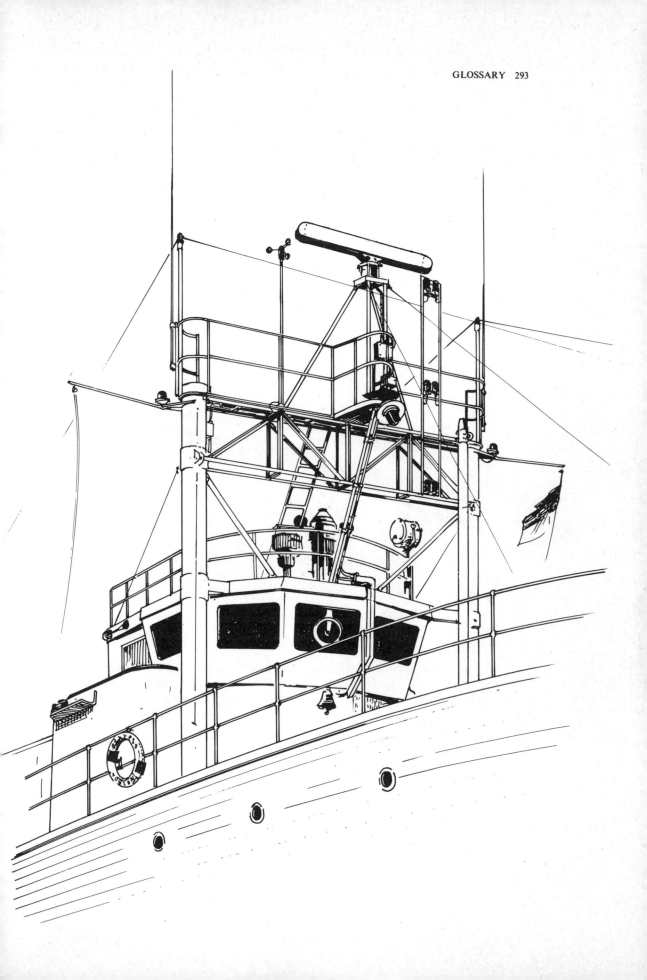

REMORA

The remora, or shark sucker, is a fish belonging to the order Perciformes and to the Echeneididae family. Its front dorsal fin has developed into a sucking disc, in which muscular flaps open to create suction.

Remoras attach themselves to any large floating animal: a whale, a shark, a sea turtle, etc., and they feed on the leftovers of their host.

The largest remora reaches a length of about three feet.

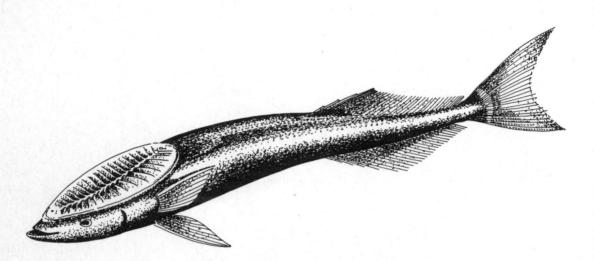

A remora.

SEA FANS

Sea fans are Cnidaria and belong to the subclass Octocorallia and the order of Gorgonacea. It takes its popular name from its fan-shaped branches and is found in yellow, mauve, and rose pink. The branches are actually animal colonies composed of polyps spread over a flexible limestone skeleton. Sea fans are fixed to the ocean floor or to rocks by branches, or tufts, and are sometimes found in tight groups. They have an encrusting base, which allows them to remain attached to their support.

Sea fans are found in all warm or temperate seas. In many tropical regions, they attain a height of over three feet and are one of the most beautiful features of the underwater décor.

The abundance and diversity of sea fans were first revealed with the advent of scuba diving. Our minisubs have also enabled us to discover true forests of sea fans in the depths of the Red Sea.

SEA LION

The sea lion is a marine mammal of the Otaridae family, order Pinnipedia. It is an amphibious carnivore whose body has adapted perfectly to swimming.

The most common species in the Galápagos is the California sea lion (*Zalophus californianus*).

These Pinnipedes live in a hierarchical society and are grouped into harems comprising from 5 to 20 individuals.

There are also furred sea lions in the Galápagos (*Arctocephalus galapagoensis*), a species which narrowly escaped extinction in the nineteenth century at the hands of hunters. The fact that their numbers were greatly reduced seems to have been the chief factor in their survival: they were simply too few to be worth the hunting.

Today, there are probably about 500 specimens surviving, mostly on the coast of Isabela Island.

SEA OTTER

A mammal of the family Mustelidae, the sea otter (*Enhydra lutris*) lives along the Pacific coast of the northern United States. It feeds on mollusks and breaks the shell of its victim against its chest with the help of a stone.

The otter may reach a length of over six feet. It is well adapted to life in the sea. Its rear limbs act as true fins.

A sea otter floating in its favorite position: on its back.

The sea otter rears and feeds it young in a stretch of kelp, the long algae of the Pacific, and envelops itself in kelp when it sleeps.

Otter fur is highly prized, and, for this reason, the animal has been hunted to the point that it is in danger of becoming extinct.

SLING

A length of cable or chain used to raise a heavy object. At one of its extremities, the sling has an eye through which the other end is passed, thus forming a loop to encircle the object to be raised. The sling is then pulled tight by the weight of the object when it is raised by a hoist or pulley block.

SLOPE

The line of greatest slant of a lode or of a sedimentary layer.

SPIROGRAPHIS or FEATHER-DUSTER WORM

The feather-duster worm of the Indo-Pacific is an annelid, or segmented, sea worm of the class Polychaeta. It is a sedentary creature whose tubelike body protrudes from the sandy bottom and is crowned with a colorful, featherlike crest. These "feathers" serve as gills, and also trap the minute organisms on which the worm lives. When the animal is disturbed, the feathers are drawn into the tube and down its spiral.

SPONGES

Sponges are the simplest of the multicellular animals. Their bodies are covered with holes through which water is drawn. Food and oxygen are then extracted from this water.

Sponges are found in a variety of shapes (fans, fingers, spheres, etc.) and colors (blue, yellow, red). There are sponges with horny skeletons (such as the bath sponge), and others have harder skeletons made of silica.

STAGES

A pause which a scuba diver must observe in his ascent to the surface in order to prevent the formation of air bubbles in his blood stream from the gas which has formed as the result of pressure.

Tables have been compiled to show the number and duration of stages necessary after a certain length of time spent at a certain depth.

STAGHORN CORAL

The common name of a madreporarian of the order Scleractinia, suborder *Astrocoeniina*. This is a madrepore of the genus *Acropora*, the form of which is evocative of the horns of a stag.

STALACTITE

A pendant deposit of calcium carbonate, resembling an icicle, formed in caverns by the accumulations of calcium salts contained in water dripping from the vault.

STALAGMITE

A deposit, more or less like an inverted stalactite, formed by the drip of calcareous water from the ceiling to the floor of a cavern.

TICK

A wingless insect which lives as a parasite on dogs, cattle, sheep, and other animals, and feeds by sucking the blood of the host animal.

TORTOISE

The giant tortoise of the Galápagos, *Testudo elephantopus*, sometimes attains a weight of over 500 pounds. It is a typical instance of the gigantism which seems common to insular life — the same instance which is found in the Seychelles Islands.

This land animal's shell varies from island to island in the Galápagos and according to the kind of food which the tortoise eats. It has been modified especially according to whether the animal feeds on a local herbaceous plant (*Paspalum conjugatum*), or on opuntia.

The tortoises of each island of the archipelago have adapted to local conditions.

ULTRASONIC TELEPHONE

A communication device which makes use of the same waves as those of sound, but the frequencies of which are beyond the limits of audibility by humans.

UNDERSEA HOUSES

Captain Cousteau's first experiment with undersea houses (Conshelf I) took place in the Mediterreanean, off Marseilles, in 1962, where two divers remained at a depth of thirty-five feet for eight days.

The second experiment (Conshelf II) was in the Red Sea, at Shab Rumi, in 1963. There, two oceanauts lived for a week at 80 feet, and eight others lived for a month at 37 feet.

The latest experiment was in 1965, and was called Conshelf III. On that occasion, six divers remained at over 300 feet for three weeks in an undersea house built in the open sea off Cape Ferrat.

UNDERWATER SCOOTER

The underwater scooter is a new, electrically powered underwater vehicle which can travel at two knots for approximately fifteen minutes. It was designed and built by the Center for Advanced Marine Studies of Marseilles.

Its plastic hull contains, in addition to its engine, a double air tank and a breathing apparatus for the diver's use.

WET SUBMARINE

A miniature submarine in which the diver remains in contact with the water and receives air from four tanks and a breathing apparatus carried in the vehicle.

The wet submarine is powered by two engines and two propellers and has three buoyant elements. The hull is of polyester, and the forward section is a transparent dome which is put in place and removed while the vehicle is in the water.

WALK

An oceanographic "walk" is a platform situated near the top of a submerged cliff. Its width varies from a few inches to six feet or more. It often owes its origin to chemical erosion, but its development is largely the result of calcareous marine life, such as madreporarians and lithothamnion, which is a calcareous alga.

WINDLASS

A machine for hoisting or hauling, as by turning a crank so as to wind up a rope attached to a weight.

BIBLIOGRAPHY

Bushnell, G. H. S. *Le Pérou*. Paris, 1958.

Disselhoff, H. D. *Les grandes civilisations de l'Amérique ancienne*. Paris, 1963

Eiseley, Loren. *Darwin's Century*. New York, 1961.

Engel, F. A. *Le monde précolombien des Andes*. Paris, 1972.

Fleming, Sverdup Johnson. *The Oceans*. New York, 1942.

Flornoy, Bertrand. *L'aventure Inca*. Paris, 1955.

Francis-Boeuf, Cl. *Les Océans*. Paris, 1947.

Jastrow, Robert. *Des astres, de la vie, et des hommes*. Paris, 1972.

Johnstone, L. *A Study of the Oceans*. London, 1930.

Métraux, Alfred. *Les Incas*. Paris, 1968.

Moorehead, Alan. *Darwin and the Beagle*. New York, 1969.

Nelson, Bryan. *Galápagos Islands of Birds*. London, 1968.

Rudaux, Lucien. *La terre et son histoire*. Paris, 1947.

Thornton, Ian. *Darwin's Islands*. New York, 1971.

Villaret, Bernard. *Le Pérou mort et vif*. Paris, 1964.

Zuber, Christian. *Archipel des Galápagos*. Paris, 1961.

Index